艺术空间的转变

Transformative Art Spaces

[丹] BIG建筑事务所等 | 编

李璐 | 译

大连理工大学出版社

艺术空间的转变

Transformative Art Spaces

城市的复杂性和抽象性
Urban Complexity and Abstraction

Aldo Vanini + Silvio Carta

实体的/数字的

作为生活在城市里的人，我们往往会受到身边发生的技术变革的影响。这种现象一直都在发生：例如，想想19世纪下半叶第一个地下系统或第一个用电照明的公共街道。像往常一样，人们对这些重大变化的反应总是交织着兴奋与漠然，甚至在某些情况下是恐惧和怀疑。因此，一些人按惯例认为这些变化可能与他们没有直接关系，而另一些人则认为这些变化是全新的、积极的机遇。随着互联网（20世纪80年代末及90年代初）和数字革命（全球从模拟技术到数字技术的转变）的出现，人们对技术进步的反应发生了巨大的变化。

这主要出自于两个重要因素。

首先，这种变化发生的速度比以往任何时候都要快。例如，摩尔定律就很好地说明了这一点。摩尔定律指出，每个硅片上的晶体管数量都将以每年翻一番的速度增长。[1]虽然摩尔的预测（1965年）最近受到了质疑和反思[2]，但它仍然为我们理解数字技术的指数增长趋势提供了一个很好的衡量标准。雷·库兹韦尔用更通俗的术语进行了阐释："技术不仅仅是工具制造；而且是一个利用上一轮革新的工具去创造更强大技术的过程。"[3]

第二，新技术的飞速发展带来了新的复杂程度。近年来，许多人在自然、数学、计算机科学（例如，约翰·霍兰德和梅拉妮·米歇尔）以及城市环境（例如，迈克尔·巴提和史蒂芬·马歇尔）领域研究复杂性、应急性和自组织系统的科学。巴提在他的名为"城市科学"的网站上，针对城市的复杂性（引自罗伯特·迈耶斯的"百科全书的复杂性和系统科学"）提出了其明显的弊端："系统是由许多交互部分组成的，这些部分能够通过自

Physical / digital

As people living in cities, we tend to absorb the technological changes that occur around us. This phenomenon has always happened: think of the first underground system or the first electrically illuminated public streets in the second half of the nineteenth century, for example. As usual, people's reaction to such significant changes has been a combination of excitement, indifference and, in some cases, fear and skepticism. Some people have traditionally therefore dismissed the changes as something perhaps not directly relevant to them, whilst others have embraced them as new and positive opportunities. With the advent of the Internet (late 1980s and early 1990s) and what has been defined as the Digital Revolution (intended as the global shift from analogue to digital technologies), the ways in which people react to technological advancements have dramatically changed.

This is mainly due to two important factors.

Firstly, the speed at which such changes occur is higher than never before. A good indication of this is, for example, Moore's Law that indicates that the number of transistors per silicon chip would double every year[1]. While Moore's prediction (1965) has recently been contested and reconsidered[2], it still provides us with a good measure for understanding the exponential trend of the growth of digital technologies. Put it in more general terms by Ray Kurzweil: "technology goes beyond mere tool making; it is a process of creating ever more powerful technology using the tools from the previous round of innovation"[3].

Secondly, new and rapid technologies entail a new degree of complexity. In recent years, many have been studying the science of complexity, emergence and self-organizing systems, in nature, mathematics, computer science (e.g. by John Holland and Melanie Mitchell) as well as in the urban environment (for example by Michael Batty and Stephen Marshall). On his website "A Science of Cities", Batty includes a clear defection for urban complexity (from Robert Meyers' "Encyclopedia of Complexity and Systems Science"), where complexity is about: "systems that comprise many interacting parts with the ability to generate a new quality of collective behavior through self-organization, e.g. the spontaneous formation of temporal, spatial or functional structures. They are therefore adaptive as they evolve and may contain self-driving feedback loops. Thus, complex systems are much more than a sum of their parts. Complex systems are often characterized as having extreme sensitivity to initial conditions as well as emergent behaviors that are not readily predictable or even completely deterministic"[4]. The notion of complexity sheds new light

组织生成新的集体行为标准，例如，自发形成时间、空间或功能结构。所以，它们在进化过程中具有适应性，并可能包含自动反馈回路。因此，复杂的系统不仅仅是各部分的总和。复杂系统的特征通常是对初始条件和紧急行为具有极高的敏感性，而这些行为往往是不容易预测的，甚至是完全确定的。”[4]复杂性的概念为我们信仰牛顿学说对于城市空间和社会关系的传统观念带来了新的启示，这种观念具有强烈的决定论，即对特定的行为会产生特定的反应（或反作用）。复杂性带来了新的不可预测性和不确定性。2012年，理查德·森尼特在讨论智慧城市时，曾有过这样一段著名的描述："城市不是一台机器……这个版本的城市可以让生活在它高效便利的怀抱中的人们变得茫然和麻木。我们希望城市运转良好，但也要对现实生活中的变化、不确定性和无序持开放态度。”[5]

日益增长的复杂性和与之相关的某种程度的不确定性对我们体验城市和建筑环境的方式产生了重大影响。事实上，城市实体环境及社会环境的复杂性呈指数级增长，加之互通式的远程信息处理网络的使用日益增加，削弱了我们通过感知和表征来了解当代城市的能力。换句话说，我们再也不能仅仅依靠感官来全面体验城市环境了。这在一定程度上与数字城市被叠加到实体城市的事实有关。这两个城市维度完全分离，但又共存。数字城市实际上是一个没有精确边界的虚拟环境，它的街道和广场与支撑它的电信网络的物理基础设施并不一致，但它们与日益密集的通信、论坛、社交网络团体和博客有关。

更重要的是，数字城市是新型关系和虚拟世界相遇的场所，在这里，个人和社区的整体观念会得到重新考虑。事实上，如果网络空间（这里指的是数字环境）一开始只是将物理环境的一部分转换为虚拟环境，那么我们的社会生活现在就可以被认为是一个独立于空间维度的新的社会维度。如果从一方面来看，社会关系是新的（例如，考虑社交媒体上的朋友／联系人的想法），那么另一方面，它们也非常不同于实体接触。我们的社会生活以及城市体验，不再仅仅发生在实体空间中，在数字世界中有一些零星的扩展。准确地说，这两个维度被认为是完全独立和不同的，但却以一种非常复杂的方式共存，因此越来越难以区分纯粹的实体或数字的东西。为了从个人的角度来观察这是如何发生的，我们应该考虑我们的大脑在感知空间（实体的和虚拟的）和社会关系的形成方面是如何工作的。

on our traditional and Newtonian idea of urban space and social relationships characterized by a strong determinism, whereby to a certain action a certain response (or reaction) would follow. Complexity brings a new degree of indeterminism and uncertainty. This has been famously depicted by Richard Sennett in 2012 when discussing smart cities: "a city is not a machine [...], this version of the city can deaden and stupefy the people who live in its all-efficient embrace. We want cities that work well enough, but are open to the shifts, uncertainties, and mess which are real life"[5].

The growing complexity and a certain level of indeterminacy related to it are having a significant impact on the way in which we experience the city and the built environment. In fact, the exponential growth of the complexity of physical and social urban environments and the increasing use of interchange telematic networks has diminished our ability to understand the contemporary city by means of sensible perception and its representations. In other words, we can no longer have a full experience of the urban environment using only our senses. This is partially related to the fact that the digital city is now superimposed onto the physical city. The two urban dimensions are completely separate from each other, yet they coexist. De facto, the digital city is a virtual environment without precise boundaries, whose streets and squares do not coincide with the physical infrastructure of the telecommunication network that underpins it, but they relate to the thickening of communication, forums, social networks groups and blogs.

More relevantly, the digital city is the place of new relationships and virtual encounters, where whole ideas of individuals and communities are reconsidered. If, in fact, at its beginning, cyberspace (intended here as the digital environment) was a mere transposition of parts of the physical environment into the virtual, our social lives can now be considered as a new social dimension in its own right, detached from the spatial dimension. If, on the one hand, social relationships are new (for example, consider the idea of friends/contacts on social media), on the other they are also very different from physical encounters. Our social life, as well as the urban experience, are no longer taking place only in the physical space with some sporadic augmentations in the digital world. Rather, these two dimensions are to be considered completely separate and different, yet coexisting in a very complex manner, whereby it is increasingly difficult to distinguish between something purely physical or digital. To see how this is happening from the individual's perspective, we should consider how our brain works in relation to the perception of space (physical and artificial) and the formation of social links.

感知和社会空间性

人类的大脑并不会将物理环境（即与物理空间相关的信息）的感官知觉、记忆以及其他类型的传入信息分开。然而，由于大脑能够优化资源，并处理与基本生存机制相关的信息，所以感官知觉有一个专有的结构，确保其对信息的快速处理。这种运作方式不应与帮助中枢神经系统处理更多一般信息的其他机制相混淆。20世纪70年代，奥基夫和纳德尔[6]在大脑的海马体和内嗅皮层中识别出专有的结构，这些结构执行空间表征和情景记忆等功能。这是一次著名的发现。

经过专用的海马体结构处理而形成的嵌入式空间映像不仅决定了那些我们在空间背景下定义为纯粹被动的关系（即当我们没有太多的关注而感知到的某个地方的空间特征），也对塑造我们自己的环境产生了重大的影响（当我们更积极地去走近空间，例如，在设计领域内）。换句话说，建筑形式，同任何其他形式一样，对我们的环境进行了积极而合理的改造。它与我们的中枢神经系统的神经结构密切相关，反过来，中枢神经系统又与空间映像相关。我们的大脑作为工具，构思着形式和空间的概念，包括秩序、对称、正交和对齐，进而优化我们对周围实体空间的理解和管理，使之合理化。每一个物理空间都对应着我们大脑里的另一个新的人造空间，它的特征是每个人都有自己感知现实的方式。

城市是这种人造空间和真实空间结构最复杂的表达。因此，纵观人类的历史，城市作为由所有建造物组成的产物，其发展与我们的大脑习得的表达它的方式有着密切的关系。在某种程度上，城市一直是我们大脑所建立的社会关系网络发展的起因，同时也是其结果。这种关系直接关系到建筑和城市结构的类型学和形态学。我们的大脑将某些想法和概念投射到物理环境中，反过来，这些想法和概念又会影响我们建立社会关系的方式。

由于新数字技术的快速发展，城市和我们大脑之间的这种双重机制在最近发生了重大的变化。数字网络

Perception and social spatialities

The human brain does not separate the sensory perception of the physical environment (i.e. the information related to physical space), its memorization, and other types of incoming information. However, due to an optimization of the cerebral resources and processing of information related to prime survival mechanisms, sensory perception has a dedicated structure that guarantees its rapid processing. This operates in a way that should not be confused with the other mechanisms by which the central nervous system processes more general information. In the 1970s O'Keefe and Nadel[6] famously identified dedicated structures in the brain's hippocampal formation and the entorhinal cortex that perform functions that include spatial representation and episodic memory.

The embedded spatial maps processed by the dedicated hippocampus structure have determined not only those relationships within the spatial context that we can define as purely passive (i.e. when we simply absorb the spatial qualities of a place without paying too much attention), but they also had a significant impact on the shaping of our own environment (when we have a more active approach to space, for example within the design realm). In other words, architectural forms, as any other form of active and rational modification of our environment, are closely linked to the neural configuration of our central nervous system which is, in turn, related to spatial mapping. Formal and spatial concepts including order, symmetry, orthogonality and alignment are conceived by our brain as tools for optimizing our understanding, rationalization and governance of the physical space that surrounds us. To each physical space corresponds another version of it produced by our brain as a sort of new artificial space that is characterized by each own individual way of perceiving reality.

The city is the most complex expression of such artificial spaces and spatial constructs. As such, through the history of humankind, the city, as the result of all built objects, has been developed in close relationship with the way in which our brain has learned to represent it. In a way, the city has been the cause and, at the same time, the effect of the development of a network of social relations established by our brain. Such relations are directly related to the typology and morphology of the architectural and urban structures. Our brain projects certain ideas and notions onto the physical environment which, in turn, has an impact on the way in which we establish social relationships.

This dual mechanism between the city and our brains has changed significantly in recent times due to the fast pace of new digital technologies. The explosive spread of digital networks has progressively superimposed new unlimited and delocalized communities onto communities traditionally defined by a limited and localized physical space. The mechanisms underpinning these place-specific communities (cities) can be better understood

的爆炸性传播已经逐渐将无限制的、非本地化的新社区叠加到了传统上由有限的、本地化实体空间定义的社区上。通过对人类连接体的类比，也就是大脑区域内的可塑性神经连接映像[7]，我们可以更好地理解支撑这些特定社区（城市）的机制。

从感性体验到抽象思维

在网络空间以及城市的虚拟维度中发生的数字活动越来越多，这就要求我们所有人重新思考看待城市环境的方式。这使得抽象思维愈发重要，其特点是算法基于对环境的感知。事实上，抽象和符号思维（来自数学、微积分或量子物理）的引入，克服了经典物理学所固有的近似性，这种近似性主要局限于基于感官知觉的体验方式。例如，在当代城市环境中，实体存在的概念只有通过相信一个人可以同时在多个社交媒体上聊天和浏览网站，并同时处于一个遥远的位置（例如，在家），才能理解。我们本能地接受了一个事实，即同一个人可以同时出现在不同的地方，更重要的是，这个人也可以是不同的角色（或化身），这取决于他们所使用的社交媒体。为了做到这一点，我们的大脑进行了一系列的抽象思维（例如，空间、时间和相关的身份）。在某种程度上，抽象思维逐渐渗透到我们对社会和城市空间的日常体验中，成为我们理解周围新技术的显著方式。

因此，抽象思维变得越来越重要，不仅对我们所有的建筑环境使用者来说是如此，对那些构思和设计它的人来说也是如此。事实上，随着数字技术越来越普及[8]，设计师和规划师已经逐渐接受了它。首先通过CAM和CAD技术，然后是数据和城市分析，数字环境正在成为建筑师和规划师的标准工作立场。有趣的是，随着数字工具的使用和虚拟空间的产生，抽象思维几乎是自然而然地产生的。目前，作为处理城市认知的方法，抽象思维还没有得到广泛的认同，但应对其大力鼓励。如果我们不能完全理解实体/数字城市的复杂性，运用抽象思维无疑将为我们引领新方向。

through the analogy of the human connectome: the map of plastic neural connections within the brain areas[7].

From sensible experience to abstract thinking

The growing presence of digital activities happening in cyberspace and, by extension, in the virtual dimension of the city, requires all of us to rethink the way in which we look at our urban environment. This leads to the growing importance of abstract thinking which characterizes algorithmic and computing approaches over the sensible perception of an environment. In fact, the introduction of abstract and symbolic thinking (from mathematics, calculus or quantum physics) has allowed overcoming the intrinsic approximation of classical physics which is limited by an approach primarily based on sensory perception. For example, the notion of physical presence in the contemporary urban environment can only be grasped by believing that a person can be present in several social media chats as well as browsing websites, and being, at the same time, in a remote location (for example, at home). We instinctively accept not only the fact that the same person can be in different places at the same time but, more significantly, that that person can also be different personas (or avatars) depending on the social media they are using. In order to do this, our brain performs a number of abstractions (spatial, temporal and related to identity for example). In a way, abstract thinking gradually permeated into our daily experience of social and urban spaces becoming the obvious way to make sense of the new technologies around us.

Abstract thinking is therefore increasingly important, not only for all of us intended as users of the built environment, but also for those who conceive and design it. Designers and planners have, in fact, gradually embraced digital technologies as it became more available[8]. Through CAM, and CAD technologies at first, then later with data and urban analytics, the digital environment is becoming a standard working standpoint for architects and planners. Interestingly, abstraction is something that comes along, almost naturally, with the use of digital tools and the generation of virtual spaces. Abstract thinking is not, currently, widely considered as an explicit approach to urban perception, but it should be strongly encouraged. Doing so would assist us, if not in understanding completely the complexity of our physical/digital cities, then surely in navigating through them.

1. https://www.britannica.com/technology/Moores-law, last accessed 13 March 2020.
2. https://www.intel.co.uk/content/www/uk/en/silicon-innovations/moores-law-technology.html, last accessed 13 March 2020.
3. https://singularityhub.com/2016/03/22/technology-feels-like-its-accelerating-because-it-actually-is/, last accessed 14 March 2020.
4. http://www.complexcity.info/flows/, last accessed 10 March 2020.
5. https://www.theguardian.com/commentisfree/2012/dec/04/smart-city-rio-songdo-masdar, last accessed 13 March 2020.
6. O'Keefe J., and Nadel L., *The hippocampus as a cognitive map* (Oxford: Clarendon press, 1978).
7. Sporns O., *Discovering the Human Connectome* (Cambridge: MIT press, 2012).
8. Lynn G., *Archaeology of the Digital* (Berlin: Sternberg Press, 2013).

公共地质

Public G

Grafton Architects

在过去的20年里，位于都柏林的格拉夫顿建筑师事务所已经从一个爱尔兰国内领先的发展态势良好的小型建筑师事务所转型为世界知名的建筑师事务所。其创始人伊冯·法雷尔和谢莉·麦克纳马拉最近获得了多项大奖，包括2020年RIBA（英国皇家建筑师协会）金奖和普利兹克建筑奖。他们的作品被定义为符合现代主义传统，理解建筑形式在公民社会中所发挥的关键作用，并通过其慷慨和高尚的品格丰富了社会生活。格拉夫顿建筑师事务所的建筑通常具有形式的复杂性和空间的戏剧性，并且表达出了虚实空间之间强有力的相互作用。他们通过复杂的交通流线

Dublin's Grafton Architects have, in the last twenty years, made the transition from leading architects within a small but thriving national scene into celebrated world leaders. Their directors, Yvonne Farrell and Shelley McNamara, have recently been showered with awards, including the 2020 RIBA Gold Medal and Pritzker Prize. Their work is defined by fidelity to a modernist tradition that understands architectural form to play a key part in civil society, enriching social life through its generosity and ennobling character. Grafton buildings are defined by their formal complexity, their spatial drama, and their forceful interplay of mass and void. They programmatically charge their projects through complex circulation

Geology

图卢兹大学经济学院_Toulouse School of Economics / Grafton Architects
金斯顿大学的Town House大楼_Town House, Kingston University / Grafton Architects

公共地质_Public Geology / Douglas Murphy

设计策略、相互交叉的功能和不断变化的贯穿室内空间的视野来掌控他们的项目，经常模糊内外部空间的界限，将公共空间带入建筑深处。

在当代建筑设计非常随性的时代，格拉夫顿建筑师事务所不但深入研究了表征史和历史编纂，并且仍然致力于以人文主义的眼光去看待建筑的角色，这种视角在经历了后现代主义的批判和魅力不凡的"明星建筑"的全球品牌效应之后已经过时了。他们既不是沉默的历史主义者，也不是浮夸的唯我独尊者，他们的崛起表明了对建筑的普遍渴望，那就是以建筑的公共用途为荣。

strategies, overlapping functions and constantly shifting views across and through interiors, often blurring external and internal spaces, bringing public space deep into buildings.
In a time when much contemporary architecture is highly discursive, engaging deeply with the history of representation and historiography, Grafton remain committed to a humanist vision of architecture's role that became outmoded after both postmodernist critiques, and then the charismatic global branding of "starchitecture". Neither reticent and historicist, nor ostentatious and solipsistic, their rise suggests a general desire for architecture that is proud of its public purpose.

公共地质
Public Geology

Douglas Murphy

圣殿酒吧广场，爱尔兰都柏林，1996年
Temple Bar Square, Dublin, Ireland, 1996

在荣获2020年普利兹克奖时，格拉夫顿建筑师事务所的创始人伊冯·法雷尔和谢莉·麦克纳马拉确立了他们作为世界顶尖建筑师的地位。这家成立于1978年的公司拥有近乎于明星身份的声誉，这中间经历了一段有趣的发展。尤其是考虑到他们对"明星建筑师"虚名的明显抵制，这家公司的国际知名度在过去10年左右才开始增长。但是，在他们举办了2018年威尼斯双年展之后，他们获得了2020年的RIBA金奖，和现在这个建筑学领域的最高奖项（普利兹克奖），他们受到的尊重是毫无疑问的。

这一明星般的声誉很大程度上基于他们在世界各地的竞赛中赢得的一系列主要为学术类的项目，令人吃惊的是，这些项目将简朴的形式力量与公民的慷慨相结合，人们通常会联想到更多含蓄的建筑。格拉夫顿建筑师事务所最近设计的建筑的特点是具有复杂的体积形态、深雕的开口和孔隙、有节奏的表面连接和强烈的构造探索，包括巨大的梁和悬臂梁。虽然它们是用现代技术和当代分层施工法建造的当代建筑，但它们回顾了一个以大型混凝土建筑为主的、技术更简单的时代，其清晰的表达旨在唤起公众意识。

格拉夫顿建筑师事务所的名字来自于他们的第一个办公室的所在地——爱尔兰都柏林的格拉夫顿街。他们一直留在都柏林。这一点意义重大，因为爱尔兰共和国多数地区为农村，人口不到500万，但它却是杰出建筑人才的家园。例如，O′donnell & Tuomey、Tom dePaor和Clancy Moore等建筑师，这只是众多杰出设计师中的一小部分，他们创建了一个有创造性的生态系统，这一系统完全掩饰了在这样的环境中进行建筑实践的困难。

格拉夫顿建筑师事务所早期最重要的项目就是在这样的环境下诞生的，他们当时成立了Group 91，这是一个松散的建筑实践团体，他们一起参加设计竞赛，试图在经济不景气的时候找到工作。圣殿酒吧的总体规划用文化建筑翻新都柏

In receiving the 2020 Pritzker Prize, Yvonne Farrell and Shelley McNamara, founding directors of Grafton Architects, assured their place as leading world architects. A reputation approaching stardom is an interesting development for a practice founded in 1978 and whose international profile has only been growing in the last decade or so, especially considering their apparent resistance to "starchitect" posturing. But after their curation of the 2018 Venice Biennale, their receipt of the 2020 RIBA Gold Medal, and now the highest award in all of architecture, the respect in which they are held is without question.

This stellar reputation is largely based on a series of mainly academic projects won in competition across the world, which are startling in their combination of an austere formal force, coupled with a civic and programmatic generosity one might normally associate with far more reticent architecture. Recent Grafton buildings are defined by complex assortments of volumetric form, deeply carved openings and voids, rhythmical surface articulation, and intense tectonic explorations, including giant beams and cantilevers. And while they are contemporary buildings, built with modern technology and contemporary layered construction, they look back to an era of simpler technique, of large, articulated concrete buildings, whose clarity of expression was designed to evoke public spiritedness.

The name of Grafton Architects comes from the location of their first office on Grafton Street, in Dublin, Ireland, and they have remained in Dublin ever since. This is significant because the Republic of Ireland is a mostly rural country with a population of less than 5 million people, yet it is home to an architectural scene of remarkable talent. Architects such as O'Donnell & Tuomey, Tom dePaor and Clancy Moore, to name just a handful, have created an ecosystem of creativity that belies the difficulties of running practices in such an environment.

The earliest significant Grafton projects came out of this milieu, with the setting up of Group 91, a loose collective of practices who entered competitions together, trying to get work in economically hopeless times. The Temple Bar masterplan, renovating a dilapidated part of Dublin with cultural buildings, was the group's first

UTEC大学校园，秘鲁利马，2015年
University Campus UTEC in Lima, Peru, 2015

路易吉·博科尼大学，意大利米兰，2008年
Luigi Bocconi University in Milan, Italy, 2008

林破旧的部分，是该组织的首次成功实践。而格拉夫顿建筑师事务所自己的大楼展示了他们后来对社交活动的关注，该建筑采用了20世纪90年代的分层立面设计法，由钢结构型材精细地连接。

格拉夫顿建筑师事务所进入21世纪以来的作品向人们展示了他们对纯粹形式和复杂排列之间的张力的探索，并且使用了随时可以风化的原始材料。其委托项目包括市政建筑、国内项目和学校，但正是他们在大学项目上的经验帮助他们一跃进入国际舞台。2002年，他们在米兰路易吉·博科尼大学新经济学院的设计竞赛中获胜。

米兰的这个项目有一个简单的前提，那就是在地面有一个连续的公共空间，上层是教室和办公室，更大的空间如礼堂设置在地下，地下层通过大窗户可见。然而，要实现这一点，需要创建一个高度发达的建筑部分，通过各种方式引导光线，最重要的是，将大部分开窗朝向远离街道的方向。其设计成果具有强烈的雕塑感，悬臂结构和凸出部分拥有挑战传统的建构主义建筑雕塑的力量。

但是，当格拉夫顿建筑师事务所赢得了另一次国际设计竞赛（设计秘鲁利马的一所大学校园）时，米兰项目的空间戏剧性很快就被超越了。格拉夫顿建筑师事务所将自己的设计方案描述为"城市悬崖"，他们的参赛草图和模型，及其在2012年威尼斯双年展的展示，似乎都在显示这个项目是来自于超级建筑时代最极端的代表。这个由雷纳·班纳姆提出的野兽派实验分支，曾在20世纪60年代的一段时期内有望成为未来的建筑。

然而，起初看起来不可能的事情，实际上刚好利用了一些非常有利的建筑环境：利马拥有这样一种特定的气候，其沙漠靠近海边，几乎从不下雨，气温往往不会低于10°C，也不会高于30°C，这意味着不仅可以设计外部的循环路线，而且保温和防水也不是必需的。因此，建成的建筑以一系列的教室和教学空间为特色，这些空间被理解为漂浮在空间中的单元，一

success, and Grafton's own buildings show signs of their later formal preoccupations, channelled through a 1990s methodology of layered facades finely articulated by steel profiles.

Grafton's work into the 2000s showed them exploring the tension between pure forms and complex arrangements, with materials left raw and ready for weathering. Commissions included civic buildings, domestic projects and schools, but it was their experience in university projects that helped them make the leap into the international scene, with their 2002 competition win for a new school of economics at Luigi Bocconi University in Milan.

The Milan project has a simple premise – a continuous public space at ground level, with classrooms and offices in volumes above, and the larger spaces such as auditoria set under the ground, visible through large windows. However, achieving this involved creating a highly developed section with light channeled down in a variety of ways, and most significantly, orienting most of the fenestration away from the street. The result is intensely sculptural, and the cantilevers and projections have all the power of the most in-your-face constructivist tectonic sculpture.

But the spatial drama of the Milan project was soon exceeded when Grafton won another international competition for a new university campus in Lima, Peru. Describing their scheme as an "urban cliff", Grafton's competition drawings and models, as well as those they exhibited at the 2012 Venice Biennale, seemed to show a project derived from some of the most extreme projects of the megastructure era, that experimental offshoot of brutalism, identified by Reyner Banham, that promised for a short while in the 1960s to be the architecture of the future.

But what initially looked impossible was actually just advantage being taken of some highly propitious architectural circumstances: Lima has such a specific climate, a desert by the sea, that it almost never rains and temperatures tend to never go below 10ºC or above 30ºC, meaning that not only external circulation is

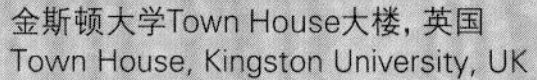
金斯顿大学Town House大楼，英国
Town House, Kingston University, UK

个巨型的翼缘墙结构沿着弯曲的场地承载着它们。贯穿其中的交通流线的设计方式让人联想起皮拉内西的风格狂野的雕塑。这面墙完全用混凝土建造，悬挂在空中花园中，在某种程度上，这是对于项目的馈赠，相当于精神上的空想社会主义。

在英国更为严谨的建筑环境中，格拉夫顿建筑师事务所最近完成了金斯顿大学的一栋建筑（36页），这是一个不太起眼的公共学习中心，但仍然探索了UTEC（利马项目）的许多相同主题。建筑师再次将令人眩晕的交通流线作为整座建筑的公共空间，但在这里，它们被包含在一个环境密封的体量中。这里显然采用了同样带肋梁的混凝土，也许更甚，因为石板的图案在墙壁和楼梯上重复出现。但在金斯顿大学项目中，在更柔和的木材和砖块的协调下，这一切都显得不那么突兀了。

外部空间和空中花园的概念仍然存在于金斯顿大楼中，但现在被理解为离开建筑时必经的露台，建筑的整个存在被一个巨大的"柱廊"连接在一起，柱廊覆盖着人工石材。这是一个网格状的巨型结构，似乎环绕着整座建筑，比例不规则，偶尔填充有露台和其他凸出结构。称它为柱廊，是将这个结构与某些市民关注的欧洲建筑联系起来。但是，它那深层网格的力量，在每层楼都用柱子微微撑起的结构的微妙姿态之下得到了凸显，这与我们可能联想到的英国建筑师彬彬有礼的设计方法几乎没有关系。

图卢兹经济学院（14页）是另一座最近完工的建筑，这次是在法国，它以另一种方式表达了这家事务所的核心主题。与金斯顿或米兰的项目相比，该建筑占据了一个更开放的场地，拥有更长的视野，它的平面布局更复杂，也更具有雕塑性，而不是一个框架体量或铰接式结构。这个理想的建筑结构一开始是由三座主要的学术大楼组成的，其中两座不规则地填

acceptable, but also that thermal insulation and waterproofing isn't required. The resulting building features a series of classrooms and teaching spaces understood as units floating in space, with a fin-wall megastructure carrying them along the curved site, through which the circulation threads in a manner reminiscent of the wilder engravings of Piranesi. Built completely in concrete, draped in hanging gardens, in a way it's a gift of a project, utopian in spirit.

In the far more restrictive architectural environment of the UK, Grafton have recently completed a building for Kingston University (p.36), a public learning centre of more humble proportions, but nevertheless exploring many of the same themes of UTEC. Again the architects have deployed vertiginous circulation routes as public spaces throughout the building, but here they are contained within an environmentally sealed volume. The same ribbed concrete is in evidence, perhaps more so, as the pattern of slabs repeats across walls and stairs, but in Kingston it becomes tempered by softer timbers and bricks.

The notion of external spaces and hanging gardens are still present in the Kingston building, but are now understood as terraces accessed by leaving the building, whose entire presence is tied together by the creation of a large "colonnade", clad in reconstituted stone. This is a gridded megastructure that appears to wrap around the entire building, irregularly proportioned, and occasionally filled with terraces and other projections. Calling it a colonnade relates the structure to certain civic concerns within European architecture, but the force of its deep grid, accentuated by a subtle gesture of the structure corbelling out slightly each storey, has little to do with the politeness that we might associate with British architects.

The Toulouse School of Economics (p.14), another recently completed building, this time in France, presents another way that the core themes of the practice are expressed. Occupying a more open site than Kingston or Milan, with longer views, the building operates more as a complex plan, sculpturally arranged, than a framed volume or articulated section. The parti begins as an arrangement of three main academic blocks, two of which crank irregularly to fill out the site, but this clarity is then studiously erased by a series of manipulations. The

图卢兹大学经济学院，法国
Toulouse School of Economics, France

满了场地，但是这种清晰度随后被一系列操作刻意地抹去了。主楼的核心成为一个空隙，位于整座建筑的正中央，楼内设有时下标志性的令人眩晕的交通流线，而楼与楼之间的连桥模糊了平面设计所暗含的简单性。该建筑几乎完全覆盖了一层坚固的红砖，格拉夫顿建筑师事务所用这种方式向当地环境表达了敬意，和其他地方一样，空白端立面与带肋梁的立面交替连接。进入建筑体量内部，没有覆盖层，给人留下了这样的印象：建筑是由单一的体量雕刻而成的。

在格拉夫顿建筑师事务所的作品中，人们很容易联想到地质的隐含意义。空间的复杂性、无窗的墙壁，巨大的悬挑和挑空空间都暗示着巨大的侵蚀和研磨的力量，凸显了他们使用当地材料作为覆层来体现建筑抽象性的习惯。虽然有些项目看似生硬地被连接起来，它们往往就像由巨大而完整的表面构成的、没有细节的结构，正是这两种效果让他们的项目产生了一种超出实际大小的感觉。这种纯粹的存在是他们作品的基本部分。许多当代的建筑从业者将绘画和构图视为他们工作的关键部分，与之不同的是，格拉夫顿建筑师事务所似乎没有特别强调他们的作品在其建筑成果之外的表现。

恪守对形式抽象的承诺，秉持建筑形式是公众团结的源泉的信念，似乎也是一种"复古"。在经历了几代人的硬理论、库哈斯式的讽刺或批评激进主义之后，所有这些都在格拉夫顿建筑师事务所的2018年威尼斯双年展上达到了顶峰。他们的主题"自由空间"谈到了建筑的"慷慨的精神和人情味"，对此普遍的批评回应是对旧式人文主义的困惑，尽管同时也赞扬了作品本身的极高质量。但在这个全球危机的时代，人们对更多社会组织的集体形式的渴望再度抬头，公共建筑的前一个时代的吸引力显而易见。格拉夫顿建筑师事务所受到的主要影响，例如，已故的勒·柯布西耶或保罗·门德斯·达·洛查的受社会影响的野兽派思想，都举例说明了这种普遍的慷慨，使他们的作品远远超过了雕塑表现主义。

core of the central block becomes a void at the very heart of the building, with the now-trademark vertiginous circulation, while bridges between blocks obscure the simplicity implied by the plan. The building is almost entirely clad in a robust red brick, Grafton's nod to the local context, and as elsewhere blank end facades alternate with ribbed facade articulations. Entering into the volume of the building and the cladding is absent, giving the impression that the building has somehow been carved from a single volume.

Across Grafton's work, geological metaphors come readily to mind. The spatial complexity, unfenestrated walls, huge overhangs and voids all suggest a sense of vast forces eroding and grinding away, accentuated by their habit of using local materials as cladding for the abstract forms of their buildings. And while some projects are relentlessly articulated, they are just as often blankly detailed with large unbroken surfaces, both of these effects giving their projects a sense of scale way beyond their actual size. This sheer presence is a fundamental part of their work, for unlike many contemporary practitioners for whom drawings and images are critical parts of their practice, Grafton seem to place no specific emphasis on the representation of their work outside of its built existence.

The commitment to formal abstraction and the continued belief in the power of architectural form as a source of public togetherness is also something that can seem "retro" after generations of hard theory, Koolhaasian irony or critical activism, all of which came to a head at Grafton's 2018 Venice Biennale. Their theme, FREESPACE, spoke of architecture's "generosity of spirit and sense of humanity", to which a common critical response was bemusement at the old-fashioned humanism, even while praising the extremely high quality of the work itself. But in a time of global crises in which there has been a resurgence of desire for more collective forms of social organisation, the attractiveness of a previous era of public architecture is apparent. Key influences for Grafton, such as late Le Corbusier or the socially-inflected brutalism of Paolo Mendes da Rocha, provide examples of that universal generosity that makes their work far more than just sculptural expressionism.

图卢兹经济学院
Toulouse School of Economics

Grafton Architects

红砖与图卢兹中世纪城市特色长廊的结合，营造和谐新校园

A new school combines red bricks and galleries featured in the Medieval city of Toulouse to create a harmony with it

图卢兹是一座由桥梁、码头、城墙、长廊、砖、扶壁、凉爽神秘的室内、回廊和庭院组成的城市。为了打造一座充满活力的当代建筑——全新的图卢兹经济学院，这些元素被重新诠释和改造。

该建筑位于中世纪城墙上的一个缺口处，连接着旧城和新城，兼收并蓄。吸收使其成为城市的核心和心脏，吐纳令其与城市空间的气息相连。

六面"扶墙"定义了建筑的外部界限，挡住了三个建筑体块，它们移动和伸展，呼应了场地的几何形状和城市景观。庭院、露台、柱廊、走廊和交通流线区域占据了街区之间的空间。

在中心处，一个宏伟的入口大厅形成了建筑的核心，建筑内部的生活围绕着它进行。通过一条外部缓坡进入内部，建筑围绕这个戏剧般的外部空间旋转，开放的楼梯连接所有的露台，并最终构成了"天空修道院"连桥的入口框架。

在南面，大会议室和咖啡馆堆叠在一起，视觉上，它们总是与主要的中心空间及可以俯瞰城市的外部露台相连。在北部和东部，研究室的延伸部分连接着俯瞰中心空间的社交走廊，那里有两个更安静的、斜面的庭院，提供了一个平静的空间，可以看到城市的风景。

混凝土墙结构有10.7m的净跨后张板，极具表现力，由供气竖管和消防楼梯作为核心支撑在场地周边。

实体砖结构采用本地砖：它长42cm、高5cm、深28cm，用石灰砂浆与相邻的中世纪墙体相匹配。多孔砖屏风为外部防火楼梯提供了遮蔽和日光。调制砖的表皮回应了朝向，形成了通往外部消防楼梯的带孔砖屏，提供遮阳功能和深浅不一、角度各异的投影。一面折叠的墙壁就像是有人居住的日晷。

Toulouse is a city of bridges, quays and city walls, promenades, brick, buttresses, cool mysterious interiors, cloisters and courtyards. These elements are re-interpreted and transformed to make a vibrant contemporary building, a new School of Economics.

Occupying a breach in the medieval wall at the junction of the old city and the new, the building inhales and exhales. It inhales to make a core and a heart. It exhales to connect to the spaces of the city.

Six "buttresses" define the outer limit of the building and form bookends to three blocks which shift and splay to respond to

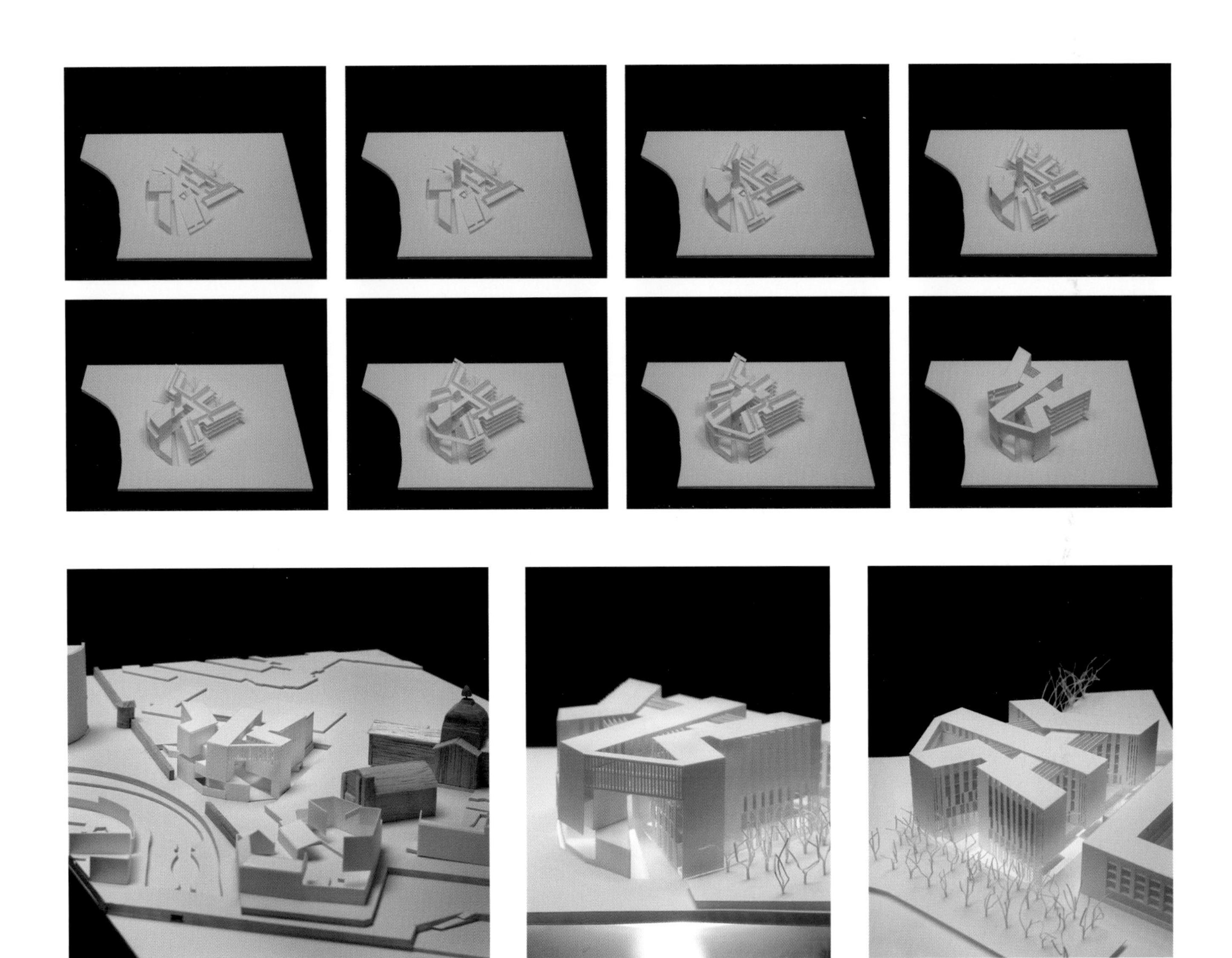

Set back.
SUNFLOWER STRUCTURE

西立面 west elevation

南立面 south elevation

项目名称：Toulouse School of Economics / 地点：Toulouse, France / 建筑师：Grafton Architects / 本地建筑师：Vigneu Zilio Architectes / 机电顾问：Chapman BDSP / 本地机电顾问：OTEIS / 工料测量师：Gleeds / 总承包商：Eiffage Construction Midi Pyrénées / 声学顾问：Gamba / 消防顾问：Vulcaneo / 客户：Université Toulouse Capitole / 总建筑面积：18,000m² 包括外部露台 / 竞赛时间：2009 / 竣工时间：2019 / 摄影师：©Frédérique Félix-Faure - p.14~15, p.17, p.18~19, p.20, p.21, p.24, p.25, p.30lower, p.31, p.32, p.33, p.34~35; ©Dennis Gilbert (courtesy of the architect) - p.28, p.30upper; courtesy of the architect - p.22

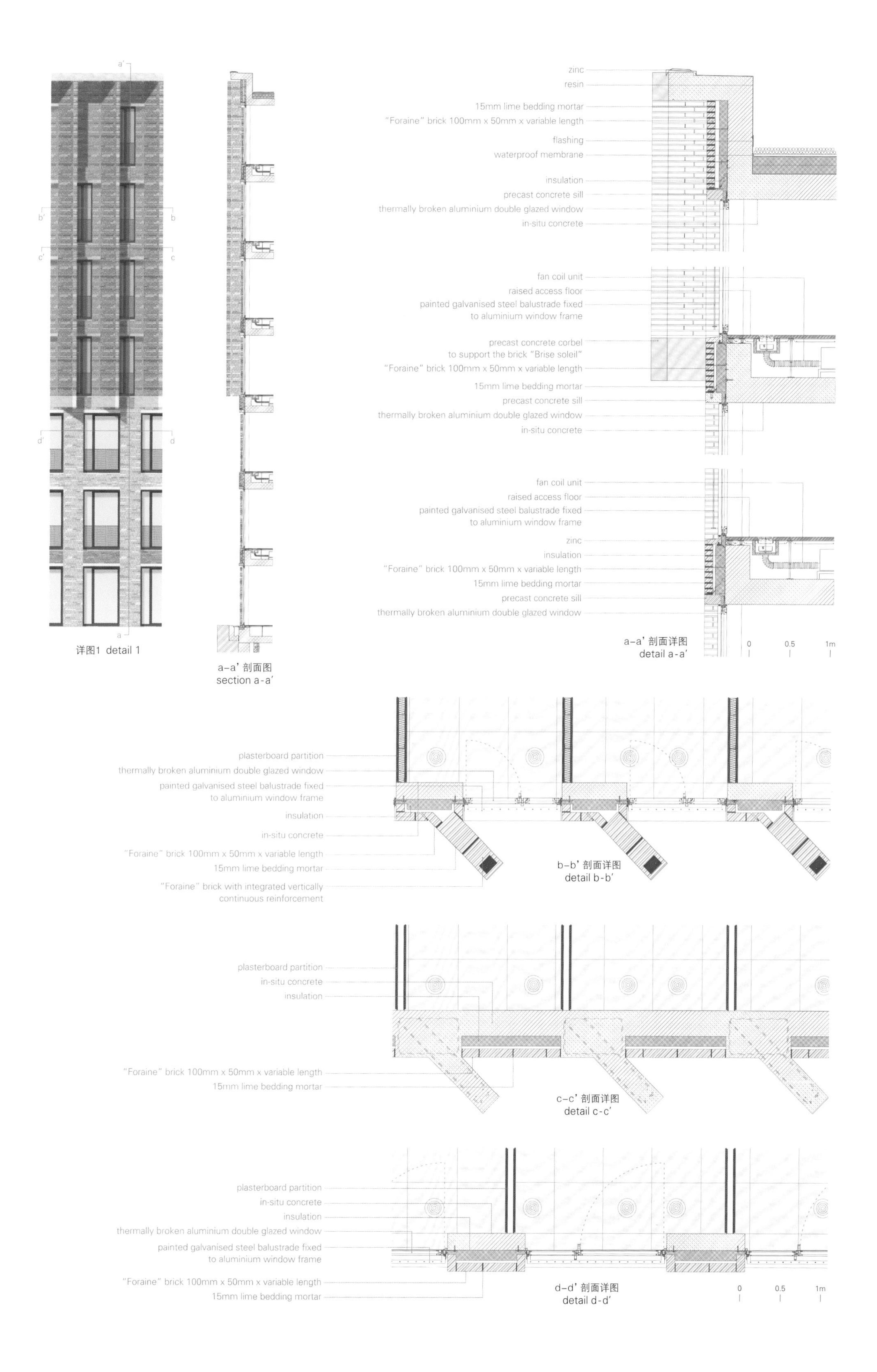

详图1 detail 1

a–a' 剖面图
section a-a'

a–a' 剖面详图
detail a-a'

b–b' 剖面详图
detail b-b'

c–c' 剖面详图
detail c-c'

d–d' 剖面详图
detail d-d'

UNIVERSITÉ
TOULOUSE
CAPITOLE
TOULOUSE SCHOOL OF ECONOMICS

一层
ground floor

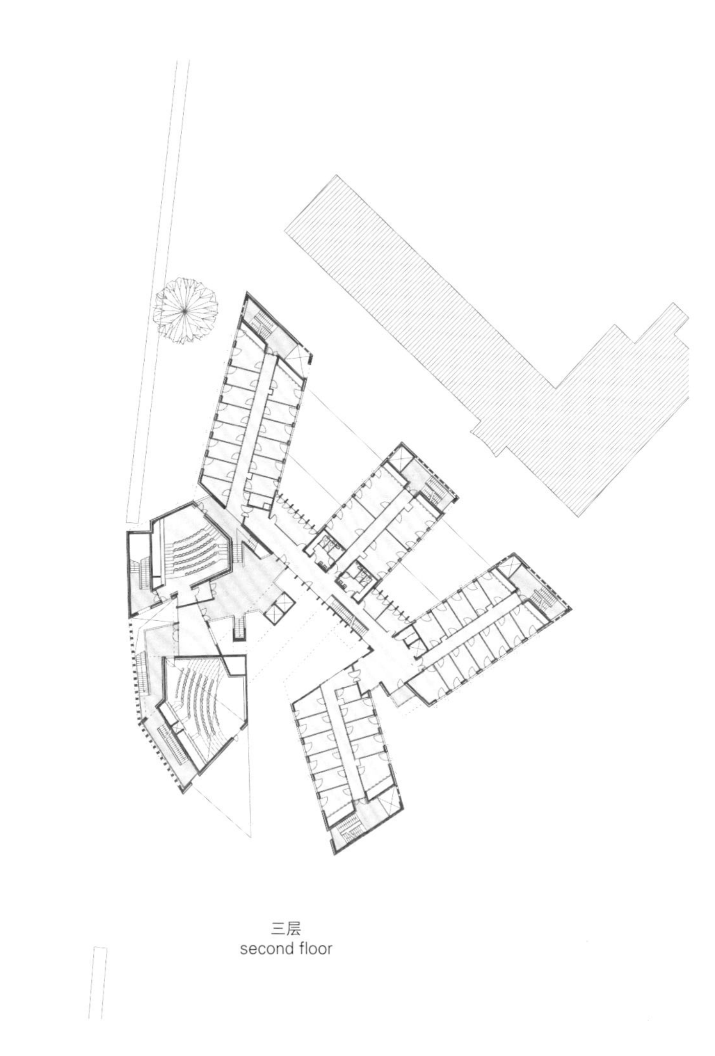

三层
second floor

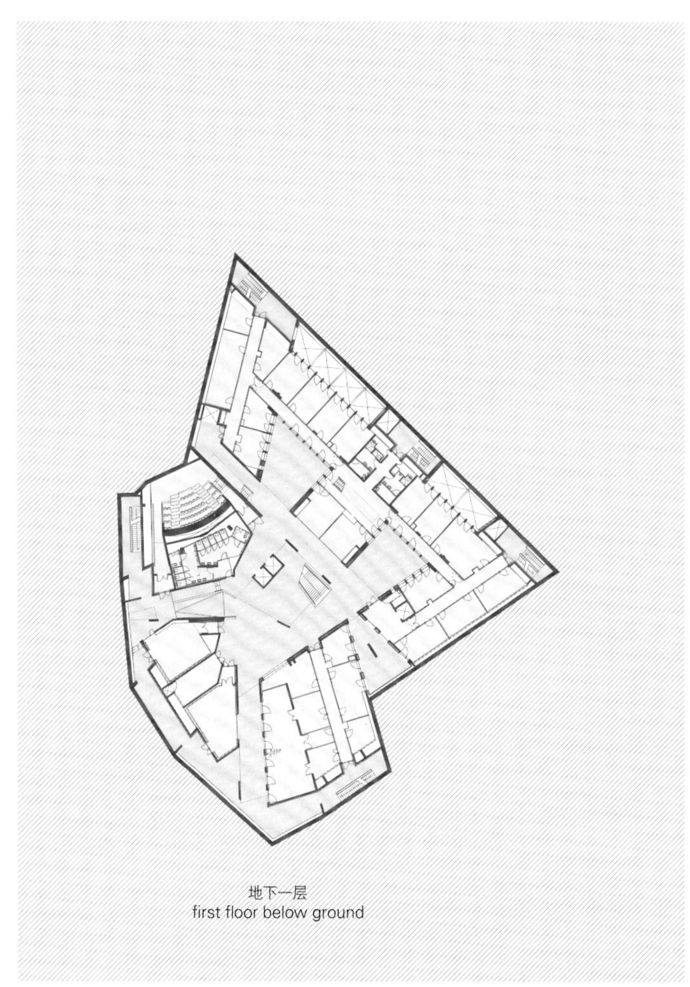

地下一层
first floor below ground

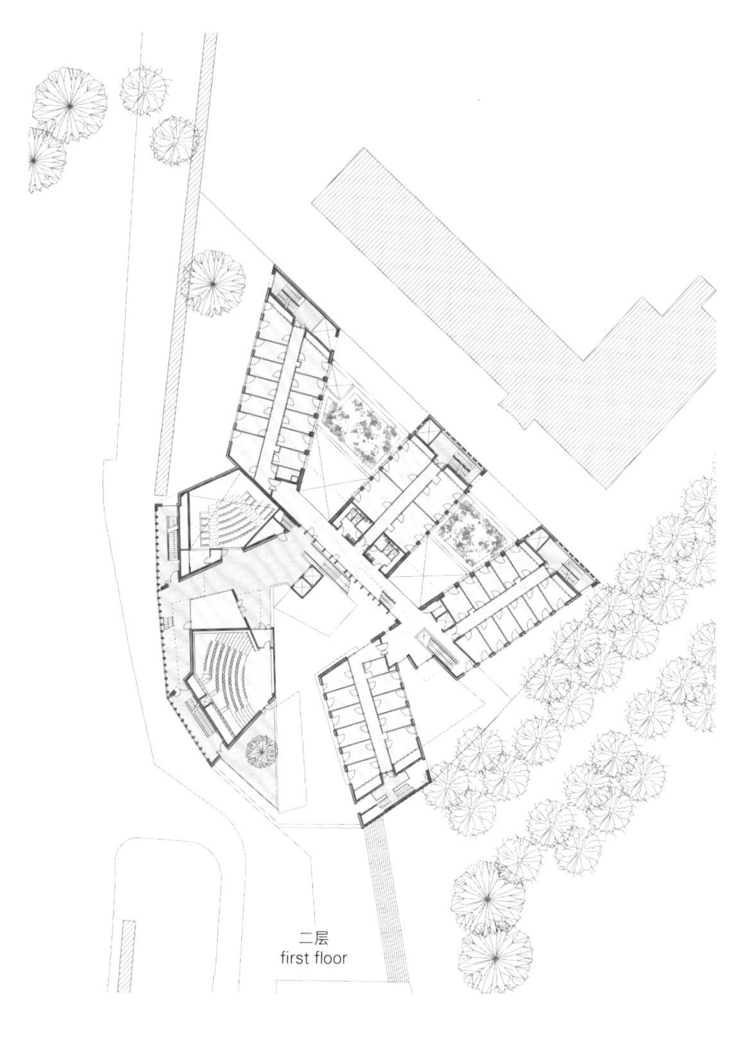

二层
first floor

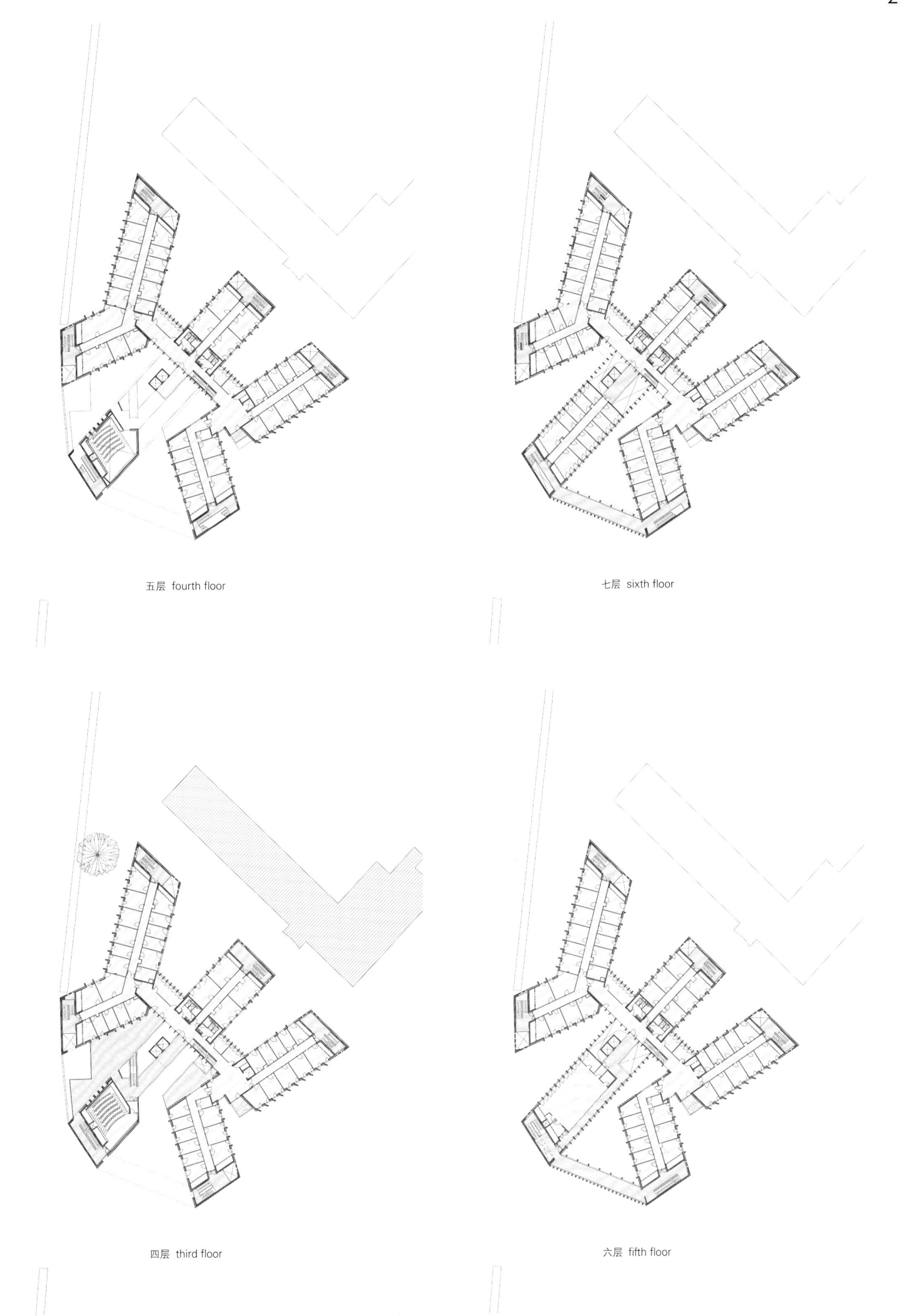

五层 fourth floor

七层 sixth floor

四层 third floor

六层 fifth floor

aile SAINT-PIERRE
accueil
bureaux T020 à T033
toilettes
aile CANAL
auditorium 3
bureaux T003 à T013
ascenseur
sortie
RdC
aile
GARONNE

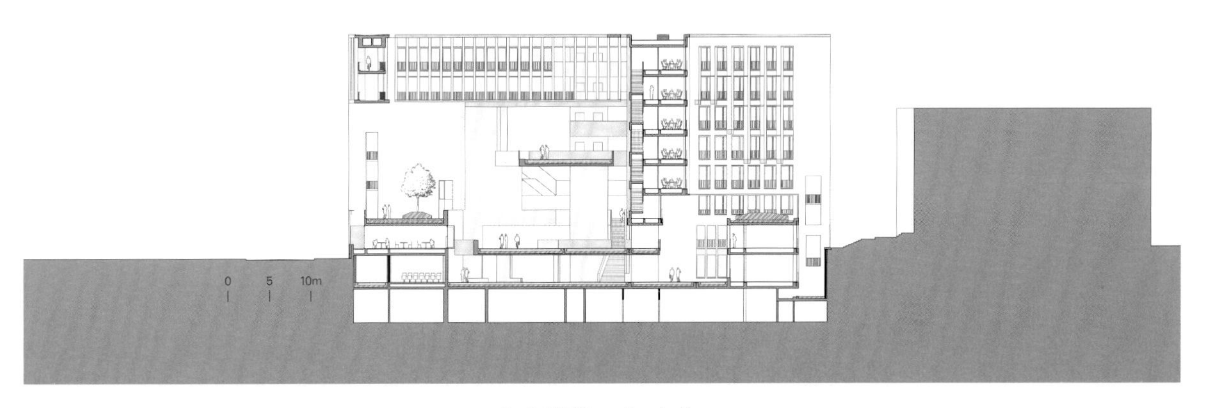

A–A' 剖面图 section A-A'

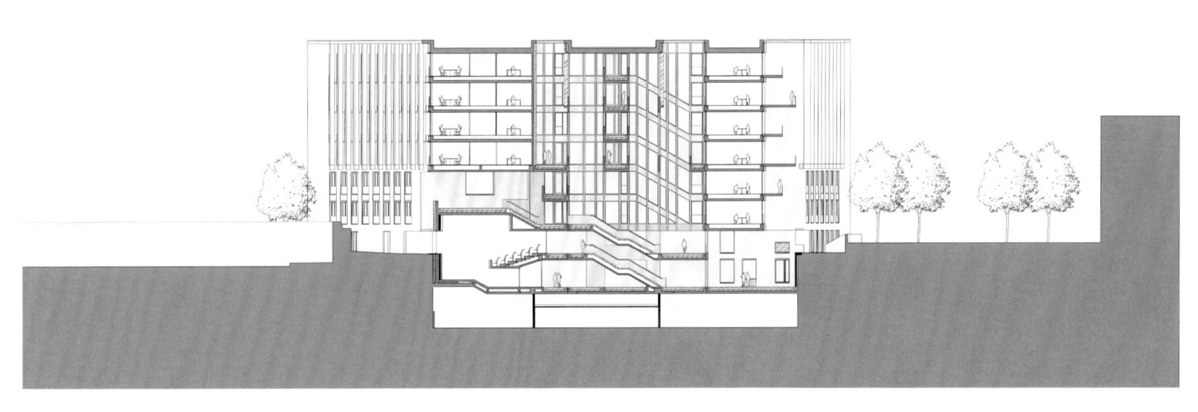

B–B' 剖面图 section B-B'

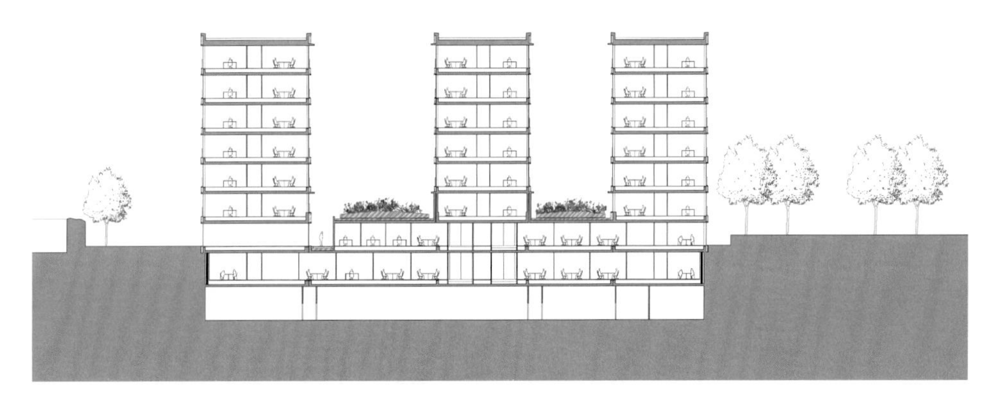

C–C' 剖面图 section C-C'

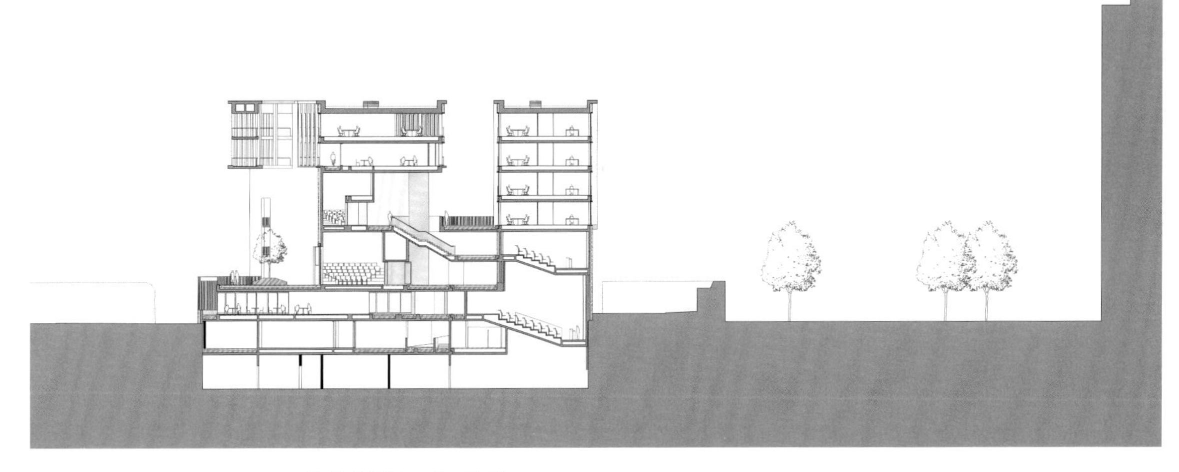

D–D' 剖面图 section D-D'

Auditorium 6

the site geometry and city views. Courts, terraces, colonnades, galleries and circulation areas occupy the spaces between the blocks.

At the center, a grand entrance hall forms the heart of the building around which the life of the building revolves. Accessed by a gentle external ramp, the building rotates around this theatrical external space with an open staircase connecting all the terraces and culminating in the "sky cloister" bridge which frames the entrance.

To the south the large seminar rooms and café are stacked, always visually connected to the main central space and to the external terraces overlooking the city. To the north and east the fingers of the research offices, connected with a social gallery overlooking the central space, with two quieter, splayed courtyards providing a calm space with oblique city views.

An expressed concrete wall structure with 10.7m clear span post-tensioned slabs is held at the site perimeter by cores of service risers and fire escape stairs.

The solid brick construction uses local bricks; it measures 42cm-long x 5cm-high x 28cm-deep, with a lime based mortar to match the adjacent medieval wall. Perforated brick screens provide shelter and daylight to the external fire stairs.

The modulated brick skin responds to the orientation, forming perforated brick screens to the external fire stairs and providing projections of varying depth and angle for sun shading. A folded wall performs like an inhabited sundial.

GARONNE
Auditorium 6
Toulouse
School of
Economics

金斯顿大学的Town House大楼

Town House, Kingston University

Grafton Architects

用柱廊和开放式平面设计打造热情的校园，构建与当地社区的联系

Welcoming Campus with colonnaded and open plan design builds a link with local community

凭借赢得了英国皇家建筑师协会（RIBA）金奖和普利兹克奖的经验，格拉夫顿建筑师事务所设计完成了伦敦金斯顿大学的地标建筑——造价5000万英镑、共六层的Town House大楼。金斯顿大学和英国皇家建筑师协会（RIBA）于2013年发起了一场设计竞赛，这家总部位于都柏林的事务所的方案入选。它是由威尔莫特·狄克逊建造的。该建筑的设计目标是成为大学的前门和通往位于泰晤士河畔的金斯顿市的门户，Town House大楼鼓励非正式学习，与市中心建立了更加牢固的联系，将活跃的学生群体与当地社区联系起来。

该建筑的开放精神在内部得到了体现，9400m²的住宿空间中50%以上设计为开放式。在实际空间中和视觉上，如同矩阵的互锁空间相互重叠交织在一起，让学生、游客和工作人员找到私密的角落进行学习和合作，同时作为整体的一部分，各矩阵之间保持相互联系。Town House大楼标志着金斯顿大学地产复兴的一个重要里程碑。它的设计考虑到了当地社区以及工作人员、学生和校友的需求，在社区内提供了一种新鲜、有活力、人们迫切需要的公民存在。正如副校长史蒂芬·斯皮尔教授所说的："格拉夫顿建筑师事务所对我们的设计大纲的诠释将确保该建筑的存在有助于消除大学和自治市之间的边界。"

柱廊是建筑的主要特征，作为北欧建筑风格中常见的外部回廊的映射，它没有障碍或层级结构，为人们提供了舒适的空间，欢迎人们进入建筑。在上层，三个层叠的景观露台形成了空中花园，从一层连接到屋顶。柱廊的外立面由加工过的石头建造，与对面市政大楼的波特兰

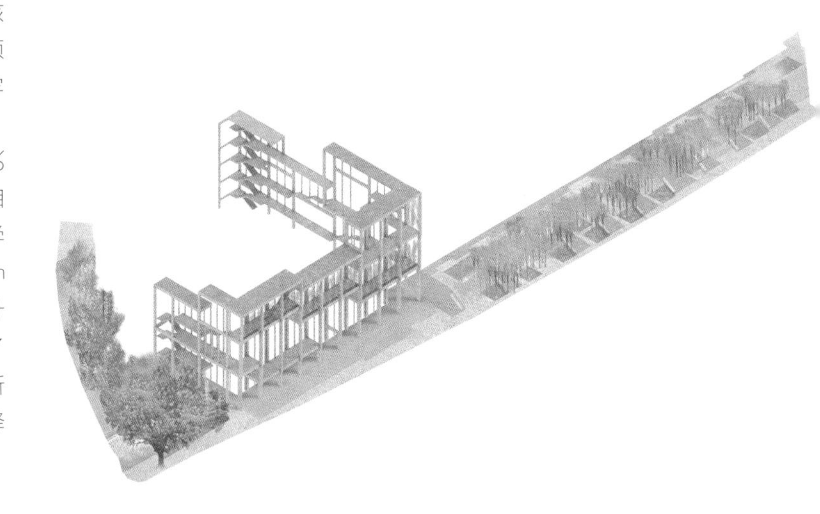

石头立面相呼应，而内立面向后设置，形成与内部空间相连的主要立面。多重入口有助于建筑的通透性。根据建筑师的说法，该建筑的设计显示了开放性。"柱廊在边缘形成了舒适热情的聚会空间。互锁的体块垂直设计，从地面连接到顶部。内部活动对路人完全可见，没有任何障碍。"

三层高的内部庭院、咖啡馆、门厅和灵活的演出空间位于一层。预制混凝土用于内部活动空间；巨大的开放式楼梯将体量和用途交织在一起；自学区域让人能够不被干扰、安静地学习。外部的柱面被优雅的内部柱反射出来，将观众的视线吸引向建筑的上层，同时落地窗让光线洒满整个空间。游客沿着楼梯移步楼上，会见到一个建筑面积最小的楼层，这里有一个咖啡馆、学习区和会议区，辅以一个室外露台，在这个露台上，从泰晤士河到汉普顿皇宫的景色尽收眼底。

带着对当地历史的欣赏，这座荣获BREEAM优秀级别认证的建筑与周围的景观相融合。经过园林绿化的新公共区域为学生和公众创造了共享空间。为了吸收水分和提高生物多样性，屋顶花园和棕色屋顶技术已经一起被纳入该方案。

RIBA Gold Medal and Pritzker Prize winning practice, Grafton Architects, has completed Town House – a £50m, six-story, landmark building for London's Kingston University. The Dublin-based practice was selected to design the scheme following a competition initiated by Kingston University and RIBA in 2013. It was constructed by Willmott Dixon. Designed to act as the university's front door and a gateway to Kingston upon Thames, Town House encourages informal learning and builds stronger links with the town center, connecting its vibrant student population with the local community.

The open ethos of the building is reflected internally, with over 50% of the 9,400m² accommodation designed to be open-plan. A matrix of interlocking spaces overlap and weave together physically and visually, allowing students, visitors and staff to find secluded corners for studying and collaborative work while remaining part of an interconnected whole. Town House signifies an important milestone in the revitalization of

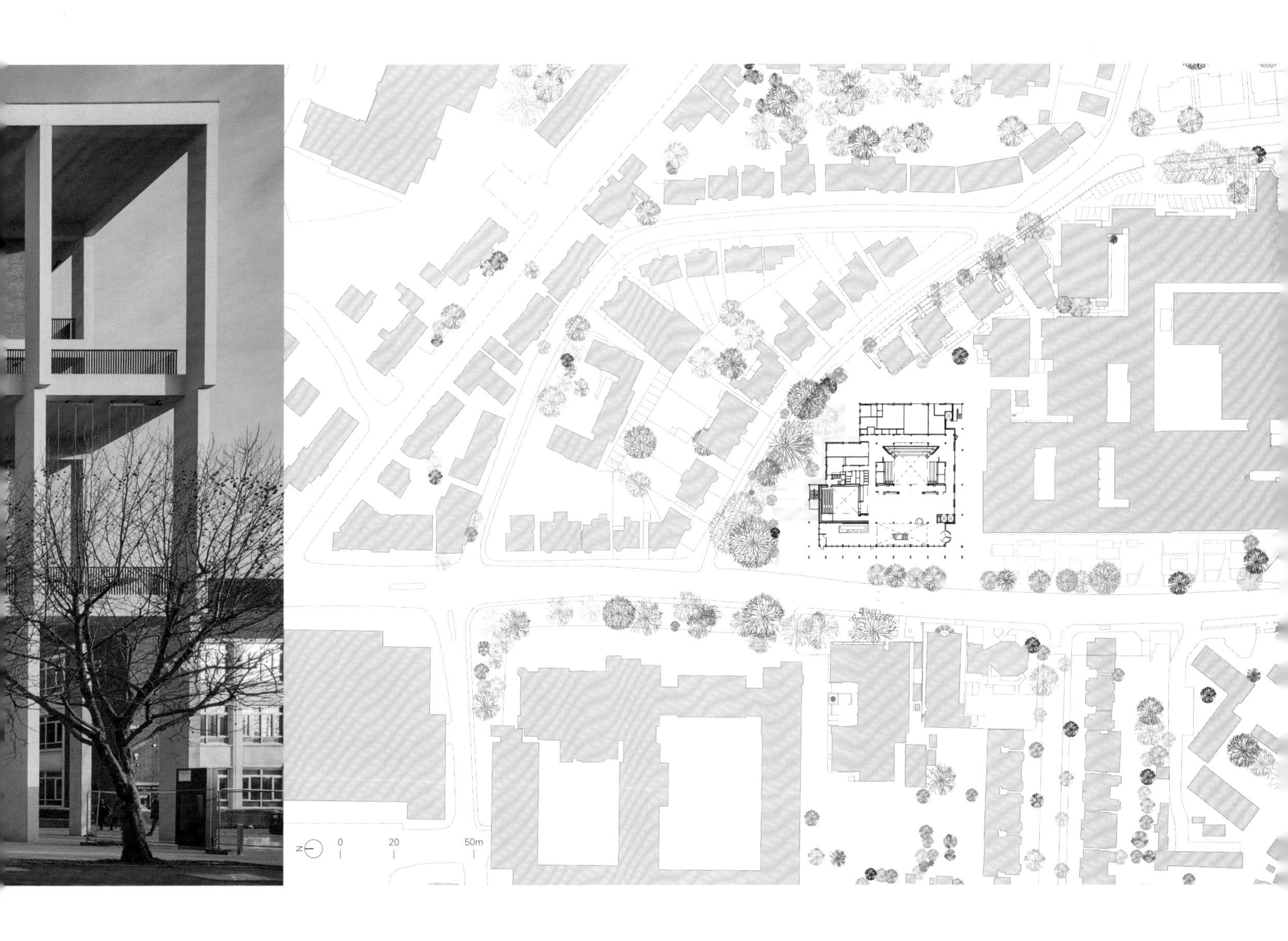

Cafe

伦敦砌砖 London stock brickwork
波特兰石材颜色的预制混凝土建筑饰面 pre-cast concrete architectural finish Portland stone color
浅灰色的预制混凝土建筑饰面 pre-cast concrete architectural finish light grey color
阳极氧化铝 anodized aluminium

南立面 south elevation

Kingston University's estate. Designed with the local community as well as with staff, students and alumni in mind, it delivers a new, dynamic, much needed civic presence in the community. As Vice-Chancellor Professor Steven Spier said: "Grafton Architects' interpretation of our design brief will ensure the building's presence helps dissolve the boundaries between the University and the Borough."

Reflective of the external cloisters seen in Northern European architecture, the colonnade is a key feature, providing amenity space with no barriers or hierarchy that is intended to welcome people into the life of the building. On the upper floors, three cascading landscaped terraces form hanging gardens, connecting the scheme from ground level to roof. Constructed from reconstituted stone, the outer face of the colonnade echoes the Portland stone facade of the municipal building opposite, while the inner facade is set back to form the main elevations which relate to the spaces within. Multiple entrances contribute to the building's porous nature.

According to the architects, the architecture reflects openness. "Colonnades form welcoming meeting spaces at edges. Interlocking volumes move vertically connecting the building from ground to top. Activities are revealed to the passer-by. There are no barriers."

The triple height internal courtyard, café, foyer and flexible performance spaces are located on the ground floor. Pre-cast concrete is used in the internal events spaces; a grand open stair weaves together volumes and uses; and private study areas allow for uninterrupted learning. The external colonnade

is mirrored by elegant internal columns, drawing the viewers' eyes towards the upper reaches of the building while floor-to-ceiling windows flood the space with light. As visitors move upwards, the stairs culminate in the smallest floor plate, which is occupied by a café, final study level and meeting space, complemented by an external terrace offering views across the River Thames to Hampton Court Palace.

The BREEAM Excellent rated building integrates with the surrounding landscape with an appreciation of local history. New landscaped public areas create communal areas for students and the public to enjoy together. Roof gardens have been incorporated into the scheme alongside brown roof technology to absorb water and enhance biodiversity.

详图1 detail 1

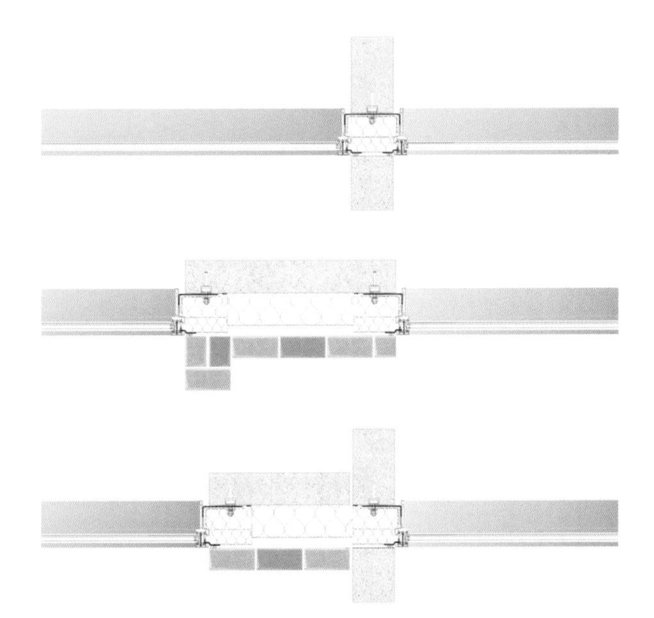

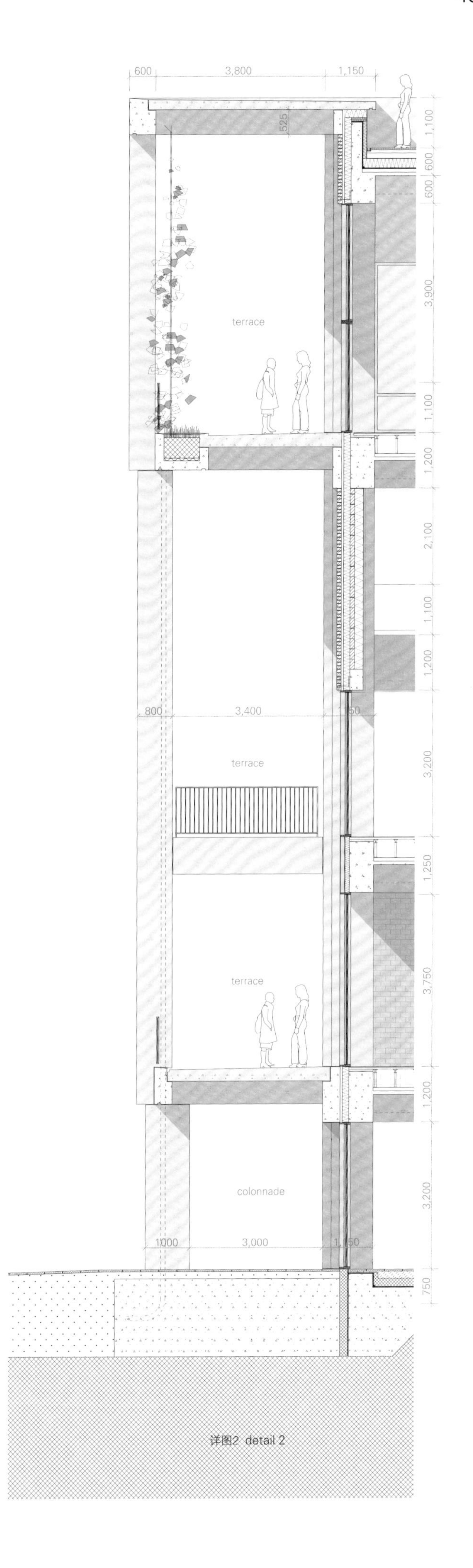

详图2 detail 2

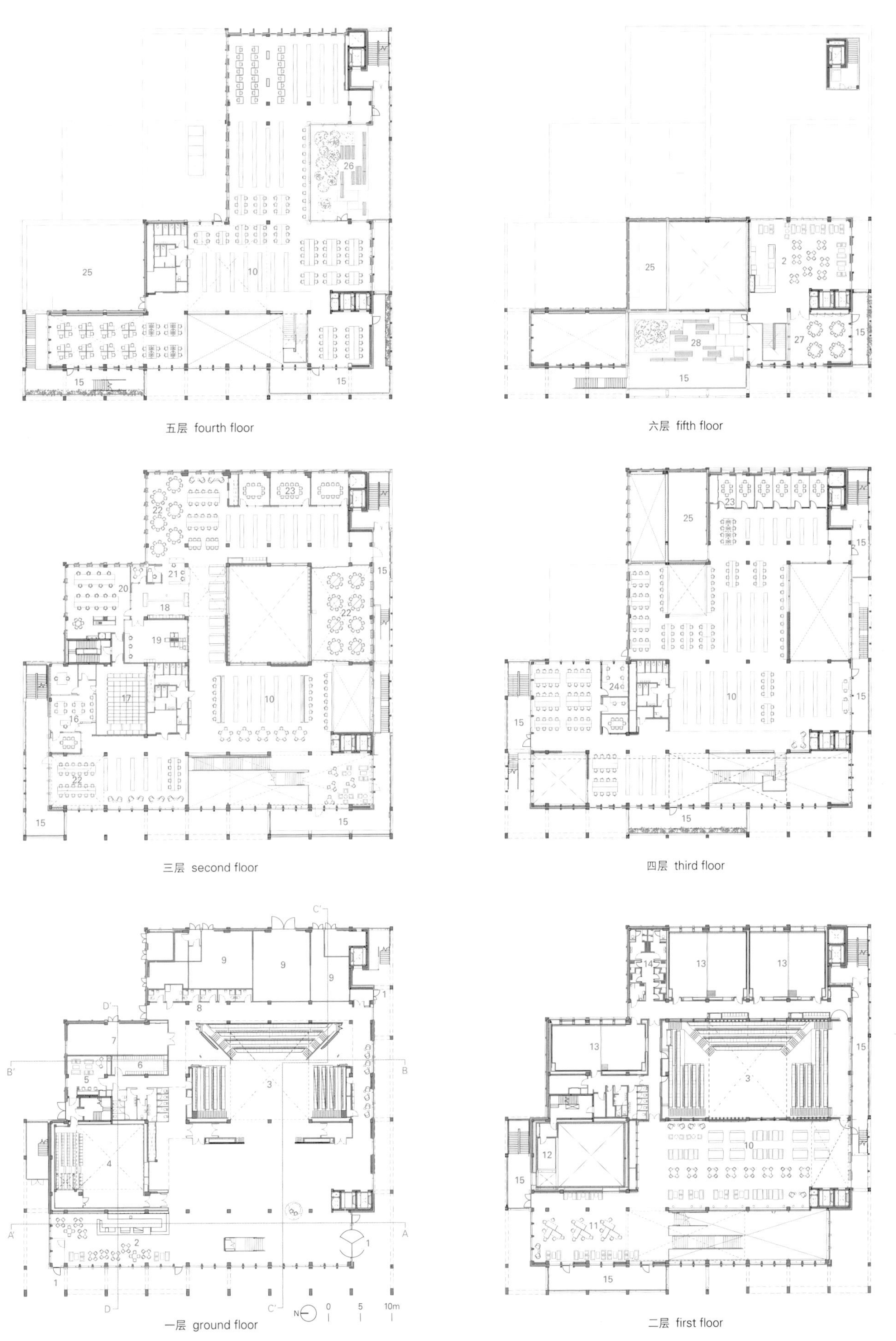

五层 fourth floor

六层 fifth floor

三层 second floor

四层 third floor

一层 ground floor

二层 first floor

1. 入口 2. 咖啡馆 3. 庭院 4. 剧院 5. 温室 6. 骑自行车者更衣室 7. 自营商店 8. 骑自行车者淋浴房 9. 设备间 10. 图书馆 11. 学生服务中心 12. 剧院控制室 13. 舞蹈室 14. 舞蹈更衣室/淋浴房 15. 室外露台 16. 特色馆藏——参考书阅览室 17. 特色馆藏——档案室 18. 复印材料区 19. 图书分类室 20. 教研室 21. 信息问询处 22. 项目空间 23. 小组活动室 24. 可访问的技术活动室 25. 屋顶绿植 26. 南侧花园 27. 角落房间 28. 西侧花园

1. entrance 2. cafe 3. courtyard 4. theater 5. green room 6. cyclist locker room 7. prop store 8. cyclist shower room 9. plant room 10. library 11. student service hub 12. theater control room 13. dance studio 14. dance changing/shower room 15. external terrace 16. special collections-reference library 17. special collections-archive 18. reprographics area 19. book sorting room 20. staff room 21. information point 22. project space 23. group room 24. accessible technology room 25. roof plant 26. south garden 27. corner room 28. west garden

项目名称：Town House, Kingston University
地点：Kingston-upon-Thames, London, United Kingdom
建筑师、执行建筑师：Grafton Architects
承包商：Willmott Dixon Construction (WDC)
结构工程师：AKT II
机电顾问：chapmanbdsp - design stage; DES Electrical / CMB Engineering - construction stage
项目经理、造价管理、工料测量、CDM&BIM咨询：Turner & Townsend
客户：Kingston University
客户技术小组：Architon Group Practice and MG Partnership
Clerk of works (Civils & Services): Fulkers
用地面积：8,535m²
建筑面积：2,385m²
室内总面积：9,100m²
总面积（包括上层的室外花园和凉廊）：10,454m²
竣工时间：2019
摄影师：
©Ed Reeve (courtesy of the architect) - p.36~37, p.38~39, p.41, p.42, p.46~47, p.48, p.49top, bottom, p.51
©Dennis Gilbert (courtesy of the architect) - p.40, p.44
©Alice Clancy (courtesy of the architect) - p.49center

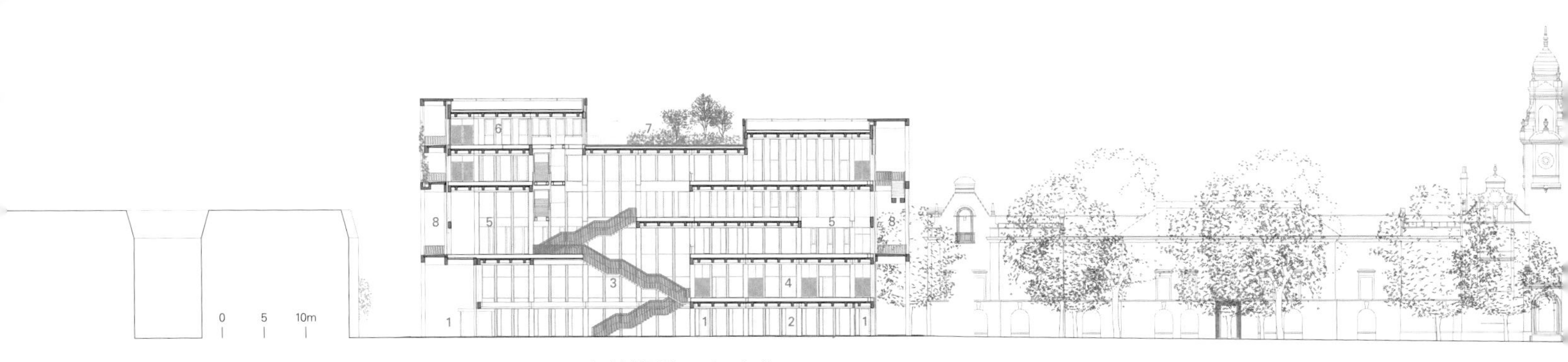

A-A' 剖面图 section A-A'

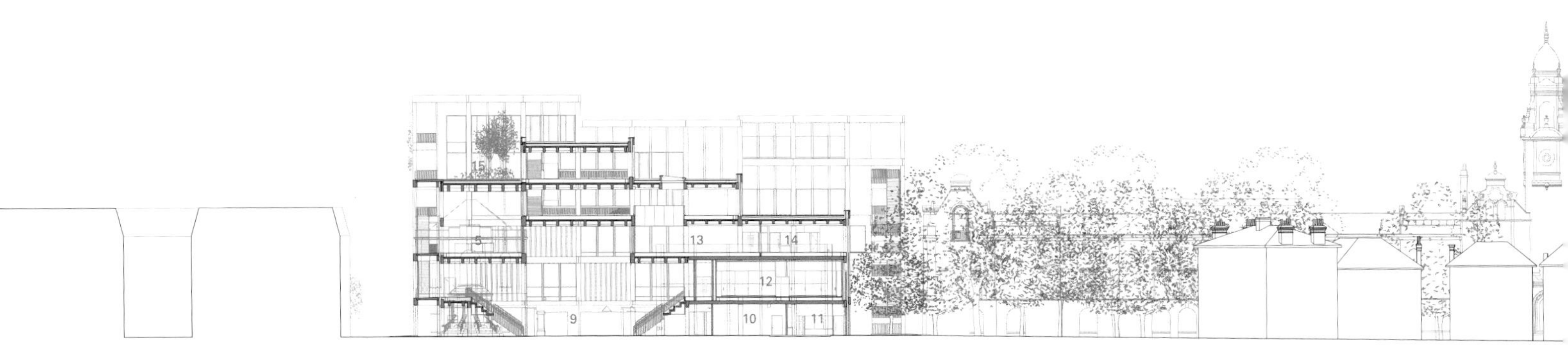

B-B' 剖面图 section B-B'

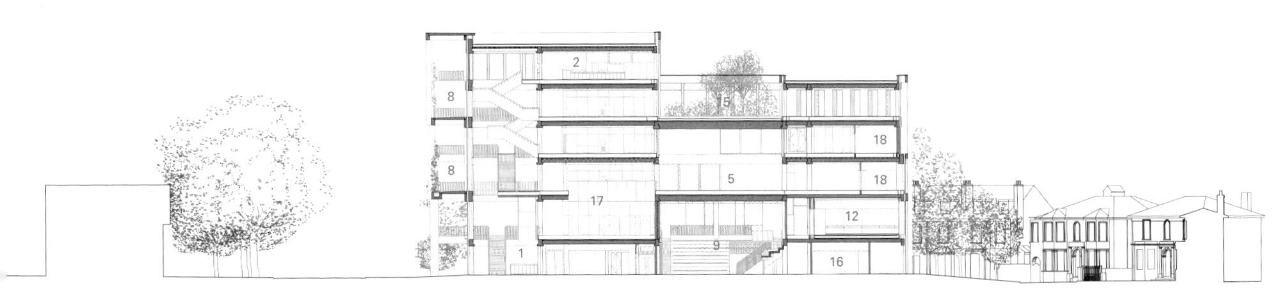

C-C' 剖面图 section C-C'

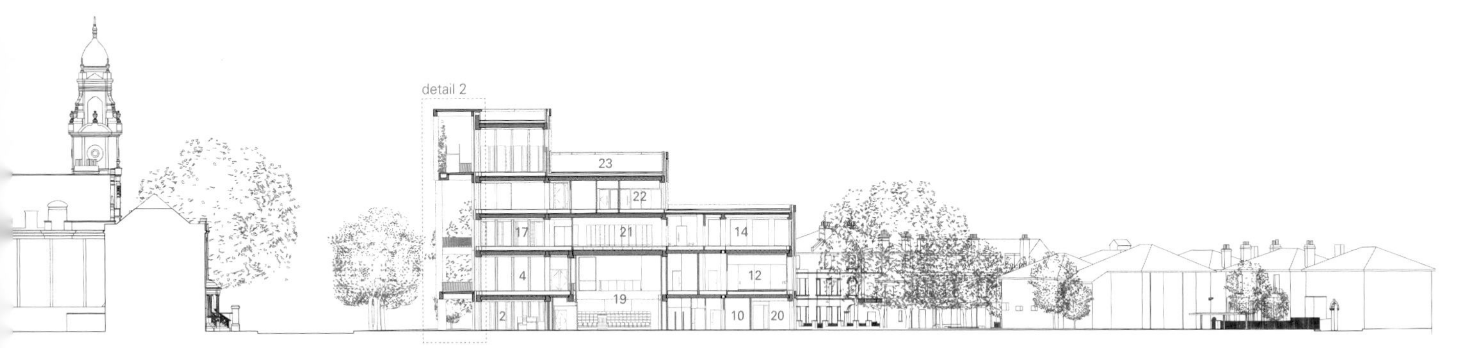

D-D' 剖面图 section D-D'

1. 入口 2. 咖啡馆 3. 中庭 4. 学生服务中心 5. 项目空间 6. 角落房间 7. 西侧花园 8. 室外露台 9. 庭院
10. 骑自行车者更衣室 11. 温室 12. 舞蹈室 13. 复印材料区域 14. 教研室 15. 南侧花园 16. 设备间
17. 图书馆 18. 小组活动室 19. 剧院 20. 自营商店 21. 特色馆藏——档案室 22. 可访问的技术活动室 23. 屋顶绿植

1. entrance 2. cafe 3. atrium 4. student service hub 5. project space 6. corner room 7. west garden 8. external terrace 9. courtyard
10. cyclist locker room 11. green room 12. dance studio 13. reprographics area 14. staff room 15. south garden 16. plant room
17. library 18. group room 19. theater 20. prop store 21. special collections-archive 22. accessible technology rooms 23. roof plant

艺术空间的转变

Transfo
Art Spa

文化驱动的城市复兴已经存在了半个世纪，并被认为是城市进入后工业环境的主要策略。文化已被视为一个关键的驱动因素，它一方面是知识经济持续创新需求的产物，另一方面也受到了消费者和体验经济的影响。一开始，这本是大城市的建设规划，但慢慢地，却渗透到了急需解决方案和艺术空间变革的城市周边地区。在这一过程中形成了一套二分法思维，即建筑项目是一个供人们消费的“打卡”文化目的地，还是一个真实的空间？这些中心是为艺术而艺术，还是为社会活动而艺术？它们应该

Culture driven urban regeneration has been around for half a century and been accepted as a primary strategy to approach our post-industrial environments. Culture has come to be seen as a key driver, as a sub-set of both the knowledge economy and its need for continuing innovation on the one hand, and the consumer, experience economy on the other. What started of as a game plan for major cities, slowly but certainly found its way to peripheral areas in need of their own solutions and transformative art spaces. A set of dichotomies have shaped within this process, and questions have risen whether architecture is employed to produce "must-see" cultural destinations for consumption, rather than for real places to be. Should these centers cater

将访客视为观众还是参与者？在这一章中，我们将仔细研究四个艺术中心，它们试图超越这些二分法，将建筑融入一个更加复杂的、能体现文化价值的环境里。这些建筑给周边地区带来了活力，寻求在经济动力、形象和社区建设之间提供一种微妙的平衡。他们认为艺术中心既是一个提供艺术展示的内部空间，也是一个延伸到城市转型和重新定位的外部空间。他们以对形状和形式的敏感度来探索这个问题，并不断地寻找内外部之间的新条件。

arts for art's sake, or arts for social functions? And should they treat their visitors as an audience or as participants? In this chapter, we will take a close look at four art centers that have sought to move beyond those dichotomies and embrace a much more complex environment in which the value of culture emerges. They invigorate the periphery and seek to provide a delicate balance between economic impetus, image and community building. They acknowledge the art center both as an interior, for the provision of arts, and as an exterior that stretches beyond, into the dynamics of city transformation and reorientation. They explore this with a sensitivity to shape and form, and a continuous search for new conditions between the inside and the outside.

艺术空间的转变
Transformative Art Spaces

Tom Van Malderen

几十年来，主要在大城市中，我们看到了以文化为主导的、广泛的城市复兴论述和案例。最初是欧洲和美国对城市“限制工业化”的回应，在这个过程中，文化被赋予了理解和改变“后工业城市”的重要角色，随后，这一方式传遍全球。归功于主要文化建筑的影响，以及富有创意和活力的文化街区的出现，文化驱动的城市复兴已经在新的城市创业精神中占据了重要的地位。在这种叙述中，文化将同时推动经济和城市的复兴。后工业化城市的理想概念不仅需要找到新的经济引擎，而且还需要解决社会凝聚力下降、市中心房产价值下滑和城市基础设施普遍衰退的问题。当然，这并不是城市第一次被消费主义、建筑、商业和文化的结合重新配置。从19世纪下半叶开始，类似的因素就开始巧妙地交织在遍布欧洲的百货商店的模型中。随着我们当代“体验经济”的到来，上述组合重新出现在一个新布局中，明确优先考虑艺术和文化。

为城市外围增加活力

更小的城镇也会随之实施他们自己的新文化基础设施计划，来刺激地方经济，重塑城镇形象并构建新社区，这只是时间早晚的问题。在这一章中，我们将仔细研究四个这样的文化中心，它们都出于某种原因而坐落于主要城市核心的外围；这些地区或多或少也受到了日益变化的经济活动和经济条件的影响。这四个项目都渴望以一种有创造性的空间营造方式去重新定义活动，改善人际联系，活跃公共和私人空间，更新建筑结构和街景，提高当地的商业活力，并将不同的人聚集在一起举行各种活动，相互启发。这些中心，以及它们所创造的环境，都在寻求利用周边城市的一种文化力量，在重塑人们的认同感方面扮演着重要的角色。它们为人们和社区提供了一个框架来（重新）架构他们自己。建筑不仅有潜力成为记忆和身份的来源，也成为关乎空间控制和空间意义的辩论的来源。

Patkau建筑师事务所设计的多边形画廊（74页）是一家位于北温哥华的新艺术画廊，帮助改造了一个以拖船和包装工厂而闻名的滨水区。与城市打造一个充满活力的海景社区中心的愿景相适应，这家画廊的目标是成为一个聚会的场所。这座由玻璃和

For several decades now, and predominantly within larger cities, we have seen extensive culture-led urban regeneration discourses and examples at work. Starting off as a European and US answer to the “de-industrialisation” of the city, in which culture was given a prominent role in understanding and changing “the post-industrial city”, the formula have spread around the globe. Thanks to the influence of major cultural buildings, and the emergence of creative and lively cultural quarters, the culture-driven urban (re)generation has come to take up an essential position in the new urban entrepreneurialism. A narrative developed in which culture would drive both economic and urban regeneration. The ideal concept for the post-industrial city required not only finding new economic motors, but also addressing declines in social cohesion, inner-city property values and urban infrastructure in general. Of course, it is not the first time that cities are being reconfigured by a combination of consumerism, construction, commerce and culture. Similar ingredients were cleverly interwoven in the model of the department stores that spread throughout Europe from the latter half of the 19th century. At the advent of our contemporary “experience economy”, the above mentioned combination reappeared in a new arrangement that was giving a clear priority to the arts and culture.

Invigorating the Periphery

It was only going to be a matter of time until smaller towns would follow suit and implement their own new cultural infrastructure initiatives in order to provide a stimulus for their economy, image and community. In this chapter we will take a closer look at four such centers for culture, somehow all located at the periphery, or outside, of major city cores; in areas that were nonetheless also affected by changing economic activities and conditions. All four projects aspire a form of creative placemaking to redefine activity, improve connectivity, liven up public and private spaces, refresh structures and streetscapes, enhance local business viability, and bring diverse people together to celebrate, inspire, and be inspired. These centers, and the environments they create, seek to employ a cultural force for said peripheral cities and play a significant role in (re)shaping people’s sense of identity. They provide a framework for people and communities to (re)configure themselves within, and have the potential to become sources not only of memory and identity but also of debate over the control and meanings of space itself.

Patkau Architects’ Polygon Gallery(p.74), a new art gallery in North Vancouver, helps reinventing a waterfront better known for tugboats and packing plants. Fitting in with the city’s vision of the seascape being a vibrant community

金属材料建成的地标性建筑极大地改变了景观，并将一个已经繁荣的社区变成了艺术爱好者的全新目的地。

北温哥华是一个人口刚刚超过5万人的独立自治市，与大城市温哥华隔着海湾相望。这两个项目都证明了小城镇雄心勃勃地用文化和创意来改造他们以前以工业为导向的城市结构，以实现经济增长，并且为社区增添活力。

阿恩斯贝格是北莱茵-威斯特法利亚的一个城市，距离德国科隆近90km，这个地区主要以其历史建筑和令人惊叹的徒步旅行路线而闻名，Bez+Kock Architekten建筑师事务所修复并扩建了Sauerland博物馆，使其成为南威斯特法利亚的博物馆和文化论坛（58页）。为了完成历史保护建筑"Landsberger Hof"和扩建博物馆的展览表面，建筑师大胆地将石灰华用于整个建筑体块的覆面，与周围历史和地理环境建立了引人注目的对话，并增加了一个从地势较低的街道的直接入口，将历史中心与周围环境以及新建成的社区连接起来。与上述海港地区相似的是，自从20世纪60年代的矿业危机以来，北莱茵-威斯特法利亚地区的历史中心逐渐变得空旷，需要找到对其重新定义的方法，并注入新的生命。当挪威传统的纸浆和造纸业开始衰落时，同样的需求也出现过。这就是Kistefos博物馆和雕塑公园建立在一家历史悠久的纸浆厂基础上的原因。建筑位于首都奥斯陆以北30km的地方，这使它担负着保护该区域建筑和工业遗产的使命，同时也代表了挪威及国际当代艺术的精华。最近，他们又担负了打造国际当代建筑的使命，邀请了Bjarke Ingels（BIG）建筑师事务所（88页）来满足其建造更大画廊的愿望，并在雕塑公园建造第二座桥，来迎合现代展示标准。BIG提出了将两座建筑合并在一起的想法，"the Twist"的概念由此诞生。

向毕尔巴鄂学习

20世纪90年代末，位于西班牙北部毕尔巴鄂的古根海姆博物馆在法国巴黎蓬皮杜中心建成20年后开馆，它向世人展示了由一个充满活力的地方议会委托设计的、充满想象力的博物馆是如何帮助一座城市扭转颓势的。所谓的"毕尔巴鄂效应"表明，城市可以通过一座标志性的建筑让自己出名，从那以后，许多城市、私人投资集团和建筑师都很难拒绝推广和尝试类似的模式。当Kistefos博物馆宣布举办"国际知名艺术家雕塑作品首展"并将博物馆的扩建部分打造为文化"打卡"目的地的时候，博物馆相继被众多媒体报道，包括《纽约时报》、彭博社和《每日电讯报》等，其建筑方案的效果可以清楚地在Kistefos博物馆的网站上看到。虽然博物馆的亮点还有很多，但是该机构还是试图通过拥有激进外观的建筑来引入它自己的"毕尔巴鄂效应"，建筑由一位著名建筑师设计，单是其结构本身就足以吸引观众并立即产生身份认同感。本章中重点介绍的四个区域艺术空间，均在其空间内建立了一个引人注目的建筑姿态。在多边形画廊这个案例里，这一姿态是通过在画廊的主要楼层上方制造"悬浮"的感觉和引

hub, the gallery very much aims to be a gathering place. The glass-and-metal landmark has dramatically changed the landscape, and has turned an already thriving neighbourhood into a brand-new destination for art lovers. North Vancouver is an independent municipality of just over fifty thousand people and sits across the inlet from the much large city of Vancouver. Both projects are testimony of smaller towns that ambitiously retooled parts of their former industry-oriented urban fabric with a cultural and creative presence, to have it followed by growth in economy, vibrancy and community resilience.

In Arnsberg, a city in North Rhine-Westphalia nearly 90km from Cologne, Germany, in a region known mostly for its heritage architecture and stunning hiking trails, studio Bez+Kock Architekten restored and expanded the Sauerland Museum to become the Museum and Cultural Forum of South Westphalia(p.58). To complement the historically listed "Landsberger Hof" and extend the museum's exhibition surface, the architects introduced a bold travertine-clad block to set up a compelling dialogue with its historic and geographical surrounding and add a direct entrance from a lower-lying street to connect the historic center with its surrounding and newer neighborhoods. Similar to the above mentioned harbour regions, the North Rhine-Westphalia area has seen its historic centers empty little by little since the mining crisis of the 60s and needed ways to redefine them, and inject new life. The same needs came to the fore when the traditional pulp and paper industry in Norway started to decline. This is how the Kistefos museum and sculpture park came to exist on the grounds of a historical pulp mill, a good 30km up north from the capital Oslo, and made it its mission to conserve the buildings and industrial heritage of the area whilst also celebrating the best of Norwegian and international contemporary art. Recently they added international contemporary architecture to their mission by inviting the Bjarke Ingels Group (BIG) (p.88) to address their wish for a bigger gallery building that caters for modern display standards, and a second bridge in the sculpture park. BIG came up with the idea of merging the two into one building and the concept for "The Twist" was born.

Learning from Bilbao

The opening of the Guggenheim Museum in Bilbao in northern Spain in the late 90s, twenty years after the Pompidou Center in Paris, France, showed how an imaginatively designed museum commissioned by an energetic local council can help turn a city around. The so-called "Bilbao Effect" came to signify the idea that a city could put

人注目的锯齿形屋顶实现的。它与之前的工业应用产生了共鸣，提供了极佳的光线分布，也提供了独特的标志性形象。

“The Twist”是一座视觉上看上去非常壮观的“可居住的桥”，桥中心的扭曲设计，增加了一段壮观的新旅程，建筑本身也成了Kistefos雕塑公园的一件艺术作品。南威斯特法利亚博物馆和文化论坛位于一个45°的斜坡上，作为新旧城镇之间的中介符号，它是一座引人注目的新建筑。

博物馆应该是什么，这是一个比其视觉存在更大的问题。但很明显，在我们这个时代，为了与博物馆建立良好的关系，你必须了解博物馆：其内部空间服务于艺术品的展示；其外部空间是城市整体、发展战略，也有可能是一场城市变革的一部分。在当今以视觉为主导的文化中，又怎么可能避开形象的重要性？这一定是件坏事吗？人们不能忽视的是，在20世纪，当艺术家们越来越多地开始研究他们周围的空间时，建筑师也开始越来越多地借鉴视觉艺术，并因此开始跨界进入视觉艺术的概念。这种艺术和建筑的融合在哈尔·福斯特的开创性作品“艺术–建筑综合体”中得到了充分的探索，揭示了在当代视觉文化日益全球化的世界中企业和文化的联盟。他指出，设计师受到极简主义的影响，艺术和建筑的相互作用也同样重要；就像极简主义者将艺术对象直接暴露于建筑条件下，这些建筑师在建筑的表面和形状方面获得了极简主义者的敏感性。正如所料，我们也可以在本章所介绍的项目中看到这一点。Bez+Kock Architekten建筑师事务所激发了他们的选择积极性，在立面上均匀地覆盖了天然石材，试图强调新博物馆体量的雕塑般的品质，同时反映了城市环境的两个主要角度。对于多边形画廊和它的表面，Patkau建筑师事务所应用了一种金属网的组合，金属网安装在镜面抛光不锈钢的上方，不锈钢的颜色从石蓝色转变为深绿色，就像周围的水随着天气在变化颜色一样。The Twist项目同样关注表面和形状，尽管它需要复杂的工程来建造它的扇形结构，但结果却出乎意料地清晰和合理。BIG的创始人比雅克·英格斯在一份新闻稿中说：“无论你往哪里看，你都会看到拱形结构和曲线形式，还有斐波那契螺旋和马鞍形状，但是当你仔细看的时候，你会发现所有的东西都是由直线构成的——直线的铝板和直线的木板。”

itself on the map via a piece of signature architecture, and it has been difficult for many cities, private investment groups and architects to resist promoting and experimenting with similar formulas ever since. One can clearly notice the influence of this blueprint on the Kistefos museum website when they announce "going international with the unveiling of sculptures and installations by well-known international artists" and by promoting the new museum extension as a "must-see" cultural destination, named by *the New York Times*, Bloomberg and *The Telegraph*, among others. Whilst there is obviously a lot more to the museum, the institution tries to introduce its own Bilbao Effect with a building of a radical appearance, and by a famous architect, whose structure alone attracts an audience and generates an immediate identity. All four regional art spaces highlighted in this chapter, set up a striking architectural gesture with the spaces they occupy. In the case of the Polygon, the gesture is made by "hovering" the main gallery floor on top of a glass box and its compelling sawtooth shaped roof. It resonates previous industrial applications, provides an excellent distribution of light but also offers a distinctive iconic image. The Twist is a visually spectacular "inhabitable bridge" torqued at its center, adds a spectacular new journey and sets itself up as an art piece within the Kistefos Sculpture Park. The Museum and Cultural Forum of South Westphalia is an eye-catching new building arranged over a 45 degree slope that serves as a mediating symbol between the old and the new town.

What a museum should be is a bigger question than its visual presence, but it is clear that in order to engage with a museum in our times, you have to understand the museum both as an interior, focused on the display of its collection, and as an exterior that is part of an urban ensemble, a development strategy, and possibly a city's transformation. In today's visually dominated culture, is it at all possibly to avoid the importance of image? And is this necessarily a bad thing? One cannot ignore that whilst artists increasingly started to involve the space around them over the last century, architects started to increasingly refer to visual art, and as a consequence have started to cross into concepts of visual art. This comingling of art and architecture was explored at length in Hal Foster's seminal work "The Art-Architecture Complex", to reveal an alliance of the corporate and the cultural in an increasingly globalised world of contemporary visual culture. He points out that with designers being influenced by Minimalism, the reciprocity of art and architecture is no less fundamental; just as Minimalists opened the art objects to its architectural condition, so have these architects acquired a Minimalist sensitivity to surface and shape. As might be expected we can also see this manifested in the projects presented in this chapter. Bez+Kock Architekten motivate their choice to clad the facade homogeneously with travertine as an attempt to emphasize the sculpture-like qualities of the new museum volume, whilst reflecting the two main angles of its immediate urban environment. For the Polygon Gallery and its surface, Patkau architects applied a combination of open-mesh metal mounted on top of mirror-finish stainless steel that shifts in color from slate blue to a deep green, much as the surrounding water changes color in different kinds of weather. A similar attention to surface and shape can be found in The Twist, and even though it required complex engineering to build its fanning form, the outcome appears surprisingly clear and

不断变化的视野、窗口和屏障

内部和外部之间的相互作用是对建筑的另一个巨大的影响，也是一个在更长的时期内反复出现的主题。想想保罗·希尔巴特出版的《玻璃建筑》，以及这本书对玻璃应用的预测，其中很多内容我们现在都认为是理所当然的，还有弗兰克·劳埃德·赖特的开放平面和密斯·凡·德·罗的玻璃幕墙；20世纪的建筑呈现出建筑内外关系的一系列变化。在本书中提到的四个艺术中心都延续了这一传统，并运用自己的边界转移。虽然艺术中心依旧传统地坚持将内部空间用于艺术品的展示，但我们可以看到展览空间通过玻璃屏得到了延伸，并可以浏览室外全景。对于一个艺术空间来讲，这是一种奇怪的反冥想，把人们的视线从艺术品上转移到迷人的城市景色或自然景色中。尽管如此，我们在工作中还是看到了一种灵活的方法，空间的一部分与艺术的呈现完美地协调，而其他部分则准备好回应活跃的公众参与的需求和"体验经济"的需求。在多边形画廊中，空间设计的主要功能围绕一个公共的玻璃基座，内有旋转展览、咖啡馆和商店。多边形画廊的副总监杰西卡·布沙尔解释说："这座建筑不是学术的圣殿，而是一个易于接近的社区中心。人们可以在这里待上一段时间，喝杯咖啡，打开笔记本电脑，逗留许久。这里将呈现一片繁忙的景象。"南威斯特法利亚博物馆和文化论坛在桥梁的末端呈现一片战略有利的视野，它连接了历史建筑和新扩建部分。一系列精心雕刻的三维窗户将这种关系从室内延伸到了野外。The Twist项目创造了三个不同的室内空间：一个更宽的、自然采光的画廊，可以看到周围环境的全景；一个中心的起伏空间，有一扇锥形的天窗；一个两层高的、南侧没有窗户的画廊。玻璃楼梯通向卫生间，那里另一个宽幅玻璃屏使游客更接近下面的河流，增强了对奥斯陆郊外田园诗般的林地的整体沉浸式体验。艺术机构就像一扇窗户，是建立我们与经常沉浸其中的世界联系的工具。当今时代，"体验经济"和心理学的意识渗透和改变了内外部的动态，改变了公共和私人的生活的动态，这些机构必须在思考艺术和消费艺术之间寻找一种微妙的平衡，积极转变观点并探索"折中"的理念。

rational. "Wherever you look, you see arches and curves, Fibonacci spirals and saddle shapes, but when you look closer you realize that everything is created from straight lines – straight sheets of aluminum and straight boards of wood," says Bjarke Ingels, the founding principal of BIG, in a press release.

Shifting Views, Windows and Screens

Another massive influence on architecture, and a recurring theme for an even longer period, has been the interplay between inside and outside. Thinking of Paul Scheerbart's "Glass Architecture" publication and its predictions for the applications of glass, many of which we now take for granted, or Frank Lloyd Wright's open floor plan and Mies van de Rohe's glass-curtain walls; 20th century architecture has unfolded as a series of variations on the relationship between the interior and the exterior of buildings. Each of the four presented art centers extend this tradition and apply their own shifting of boundaries. Whilst art centers consist traditionally of interior oriented spaces to hold the focus on the works of art, we see the exhibition spaces extend through glass screens and provide sweeping views of the exterior. For an art space it is oddly anti-contemplative, to draw the eye away from the artwork and out toward a mesmerizing city view or captivating nature scene. Nonetheless, we see a flexible approach at work, where parts of the spaces are perfectly attuned to the presentation of art, while others are geared up to respond to the need for a vibrant public engagement and the demands of our "experience economy". In the Polygon Gallery, the main functions are hovering above a public glass pedestal with space for rotating exhibitions, cafés and shops. "The building is not this temple to academia but rather an accessible community hub," associate director of the Polygon Gallery Jessica Bouchard explains. "It's a place for people to stay a while, get a coffee, open their laptop, linger. It will be a hive of activity." The Museum and Cultural Forum of South Westphalia includes a strategically positioned view at the end of the bridge that connects the historic building to its new extension, and a series of carefully carved three-dimensional windows to stretch the relationship from the interior to the hinterland it provides for and depends upon at the same time. The Twist creates three distinct interior spaces: a wider, naturally lit gallery with panoramic views of the surroundings, a central undulating space with a tapering skylight, and a double-height, windowless gallery at the southern side. A glass stairway leads down to the restroom area, where another full-width glass screen brings visitors even closer to the river below, enhancing the overall immersive experience of being in the idyllic woodlands outside of Oslo. Not unlike a window, art institutions are instruments that help us connect to the world in which we are constantly immersed. In an era where the "experience economy" and an awareness of psychology has permeated and transformed the dynamics of inside and outside, of public and private life, these institutions have to look for a delicate balance between contemplating and consuming art, and are stimulated to shift views and explore concepts of the "in-between".

阿恩斯贝格博物馆和文化论坛
Museum and Cultural Forum in Arnsberg

Bez+Kock Architekten

阿恩斯贝格上西区和下东区之间的文化延伸

The cultural extension mediating between the upper west and the lower east in Arnsberg

Sauerland博物馆位于历史保护建筑Landsberger Hof大楼中，在经历扩建之后成了南威斯特法利亚博物馆和文化论坛。一期工程对建造于1605年的老建筑进行了全面翻新，并重新设计了永久展区。二期工程在现有的建筑旁增建了一栋新建筑，倾斜45°角，向下通向鲁尔大街和鲁尔河。这座崭新的如雕塑一般的建筑，其首层空间低于原建筑的入口层约20m，使博物馆能够容纳该地区的顶级临时展览。

Bez+Kock Architekten建筑师事务所在2012年赢得了设计竞赛，但是他们根据客户的要求对获奖方案（现有建筑和新建筑通过地下层连接）进行了整体重新设计。新方案为博物馆赋予了一个由北向南逐步下降的台阶式体量：从现有建筑的地下一层延伸至建筑主体大展厅所在的鲁尔大街。沿着通往鲁尔大街的白色楼梯往下走，建筑的体量逐渐增大：从展览空间的上层开始，到夹层的多功能厅，再到宏伟的展厅。

现有建筑通过桥梁般的连接结构，在Landsberger Hof大楼的地下一层与扩建建筑相连，并通过三个朝着Brückenplatz广场敞开的斜切的窗户洞口将这条通道凸显出来。通道的尽头是一面通高的全景窗户，可从离地15m的高度欣赏令人惊叹的城市美景。位于Landsberger Hof大楼下方山丘上的历史悠久的"英格兰长廊"得到了保留，从连接两座建筑的桥梁下经过，并连接至博物馆较低的屋顶上方的公共全景露台。

引人注目、错落有致的新建筑还扮演着城市的媒介，将西面的"旧市场"与东面高度显著降低的鲁尔大街连接起来。通过对Landsberger Hof大楼和鲁尔大街之间的两个主要角度的利用，新的体量很自然地融入了其所在的建筑环境。同时，Landsberger Hof大楼作为城墙上历史悠久的重要宫殿，在阿恩斯贝格旧城区的轮廓上仍然占据着主导地位。新建筑的均质立面包覆着来自德国南部Gauingen的石灰华，强调了新博物馆的雕塑感。精心切割的三维窗户嵌入立面，创造出令人兴奋的内外空间关系。

在博物馆和文化论坛的设计中只使用了天然材料。整体立面的天然石材（石灰华）和展览空间的橡木地板具有可持续性，手工制作的材

1. 老建筑
2. 新建筑
3. 鲁尔大街
4. 旧市场
1. old building
2. new building
3. Ruhrstrasse street
4. Alter Markt

料极具耐久性。体量巨大的建筑和少量的开窗确保了高的热惯性，使建筑对温度的波动不敏感，从而很好地满足了博物馆建筑的要求。

The Sauerland Museum, located in the historically listed "Landsberger Hof", has been expanded to become the Museum and Cultural Forum of South Westphalia. To achieve this, the existing historical building, from 1605, was extensively renovated in an initial construction phase, and the permanent exhibition space was redesigned. Phase two then comprised the building of an extension, located on the directly adjacent lot, which sloped at a 45 degree angle leading down to Ruhrstrasse and the Ruhr River. This new sculptural construction, whose ground floor lies nearly 20m below the entrance level of the prestigious existing building, now enables the museum to house top level temporary exhibitions of interregional stature.

Bez + Kock Architekten won the competition in 2012, but at the client's request, the underground connection between the existing and new constructions, which featured in the original proposal, had to be completely redesigned.

The new design of the museum building is stepped like a grandstand from north to south in three stages from the level of the first basement of the existing building down to the Ruhrstrasse, where the main building mass of the great exhibition hall is located. On the way down through the white stairwell to the Ruhrstrasse the volume increases in stories – from the upper level at the start of the exhibition space, to the mezzanine level multi-purpose hall, down to the grand exhibition hall.

项目名称：Museum and Cultural Forum in Arnsberg / 地点：Alter Markt 24-30, 59821 Arnsberg, Germany / 建筑师：Bez + Kock Architekten Generalplaner GmbH – Martin Bez, Thorsten Kock / 竞赛团队：Tilman Rösch, Lisa Diez / 项目团队：Meredith Atkinson, Lea Keim, Antonia Hauser, Anna Piontek, Maria Dallinger, Roman Ramminger, Andrea Stegmaier / 本地施工管理：BBM Bodem Baumanagement / 结构工程：wh-p Ingenieure AG / 电气规划：GBI Gackstatter Beratende Ingenieure / 建筑设备工程：Henne & Walter Ingenieurbüro für technische Gebäudesysteme / 建筑物理：Wolfgang Sorge Ingenieurbüro für Bauphysik / 项目与展览概念：Dr. Ulrich Hermanns Ausstellung Medien Transfer GmbH / 景观设计：Wiederkehr Landschaftsarchitekten / 客户：Hochsauerlandkreis, represented by Landrat Dr. Karl Schneider, Meschede / 总建筑面积：3,533m²
总体积：16,271m³ / 施工造价：appr. 7,7 Mio. € (gross) / 竞赛时间：2012.2 / 设计开始时间：2012.3 / 施工时间：new construction – 2017.5; renovation of existing building – 2016.10 / 竣工时间：new construction – 2019.9; renovation of existing building – 2018.8 / 摄影师：©Brigida González

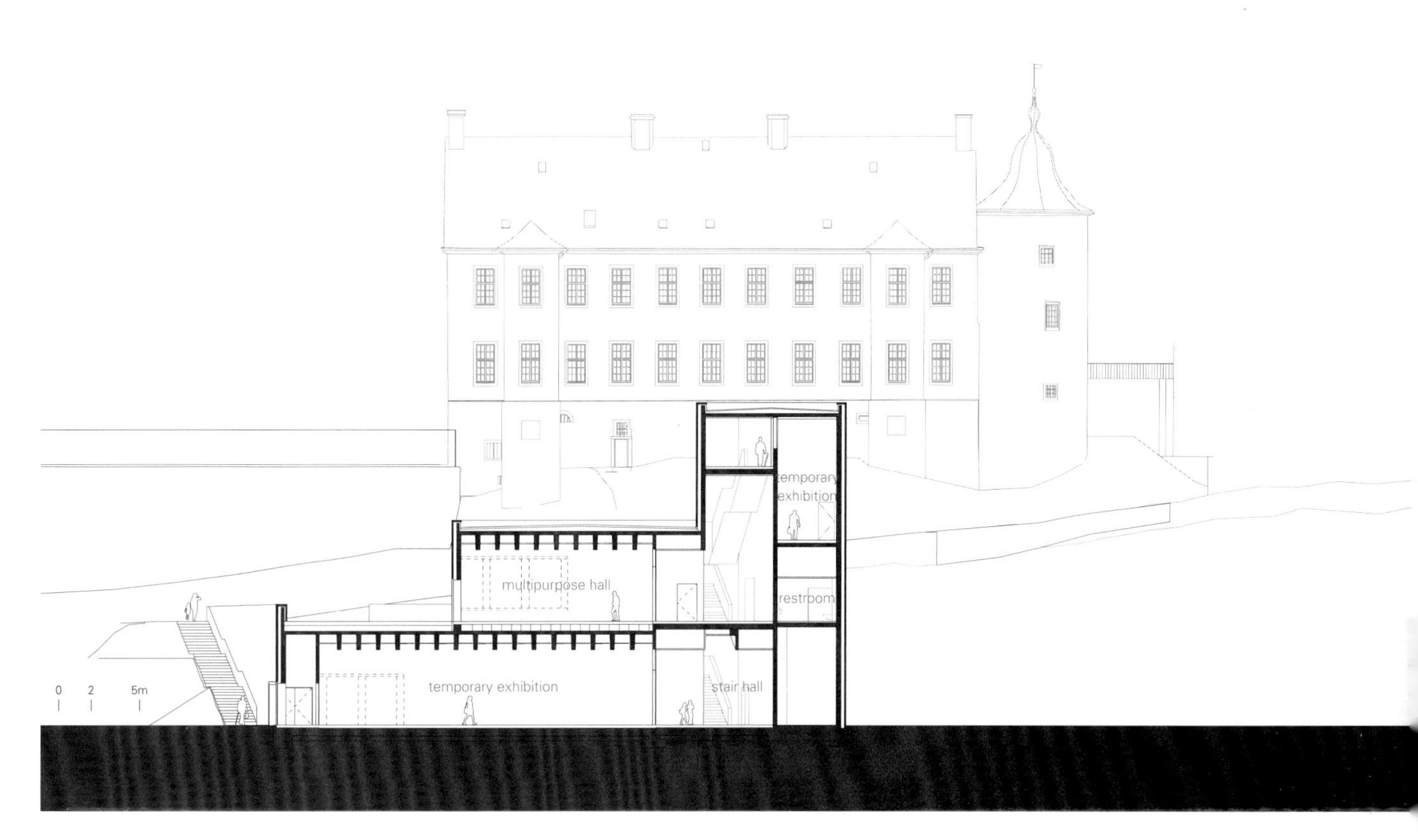

A–A' 剖面图 section A-A'

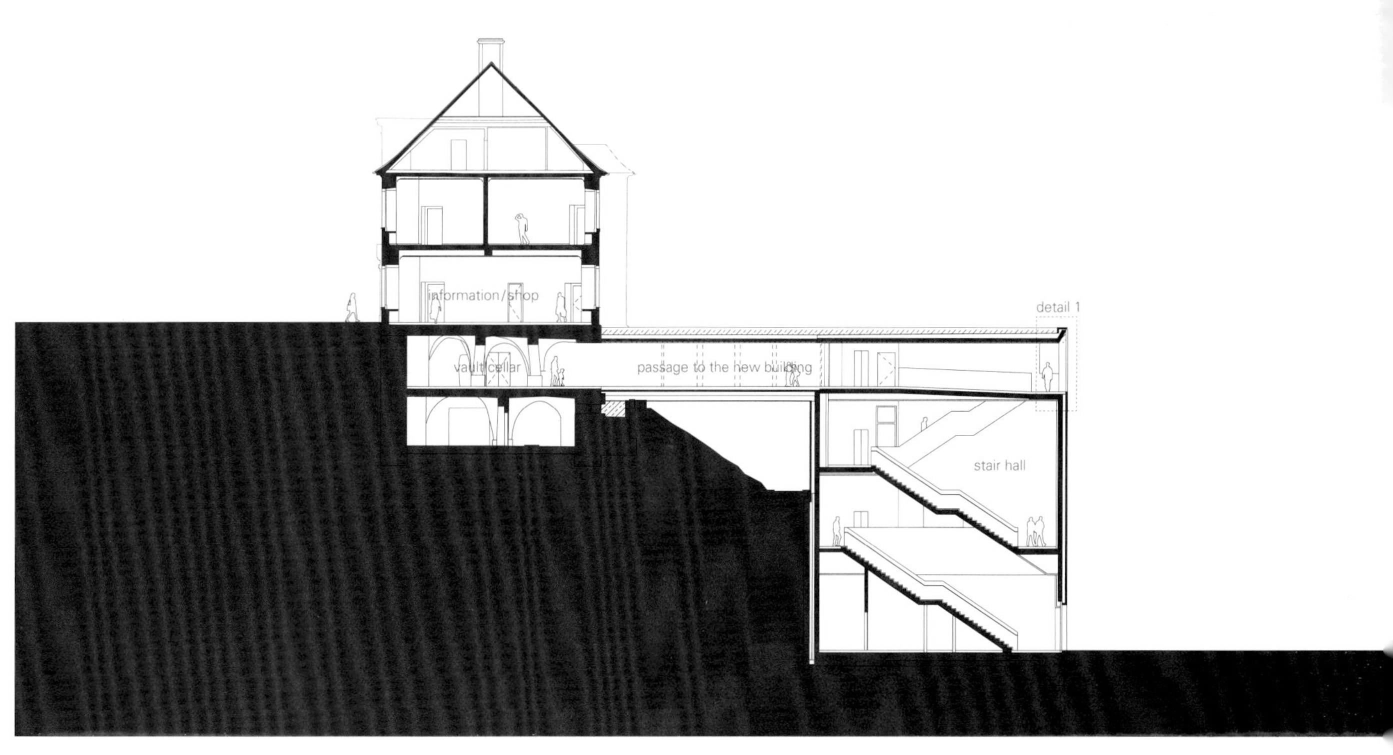

B–B' 剖面图 section B-B'

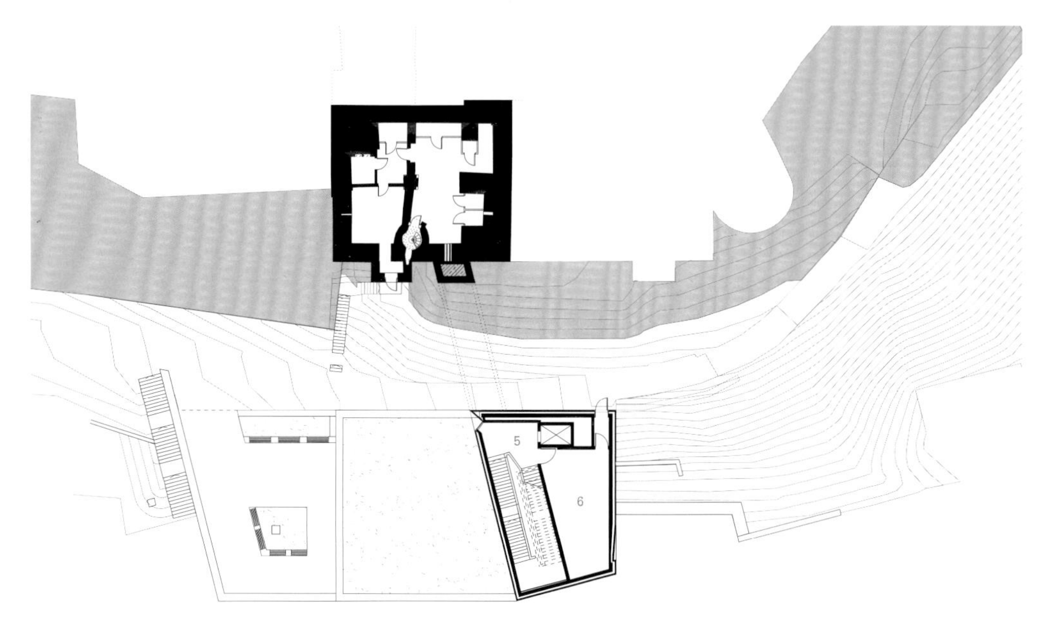

地下二层 second floor below ground

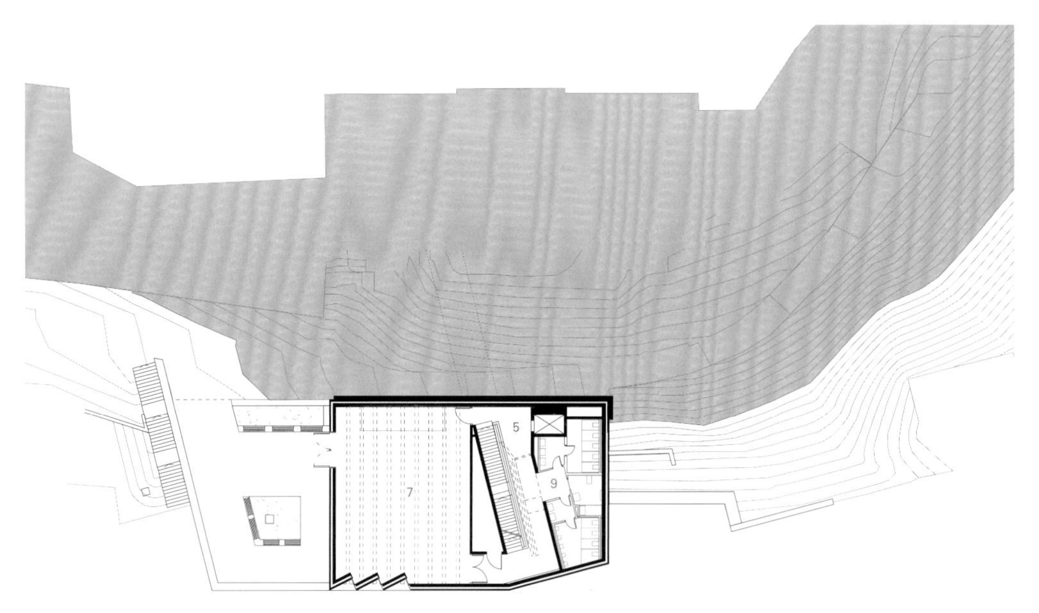

地下三层 third floor below ground

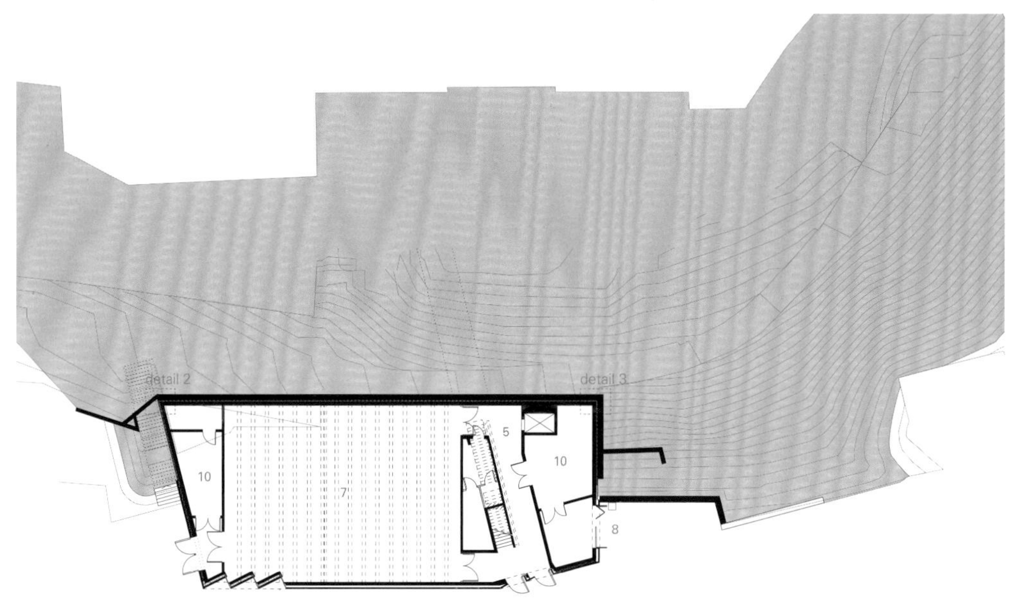

地下四层 fourth floor below ground

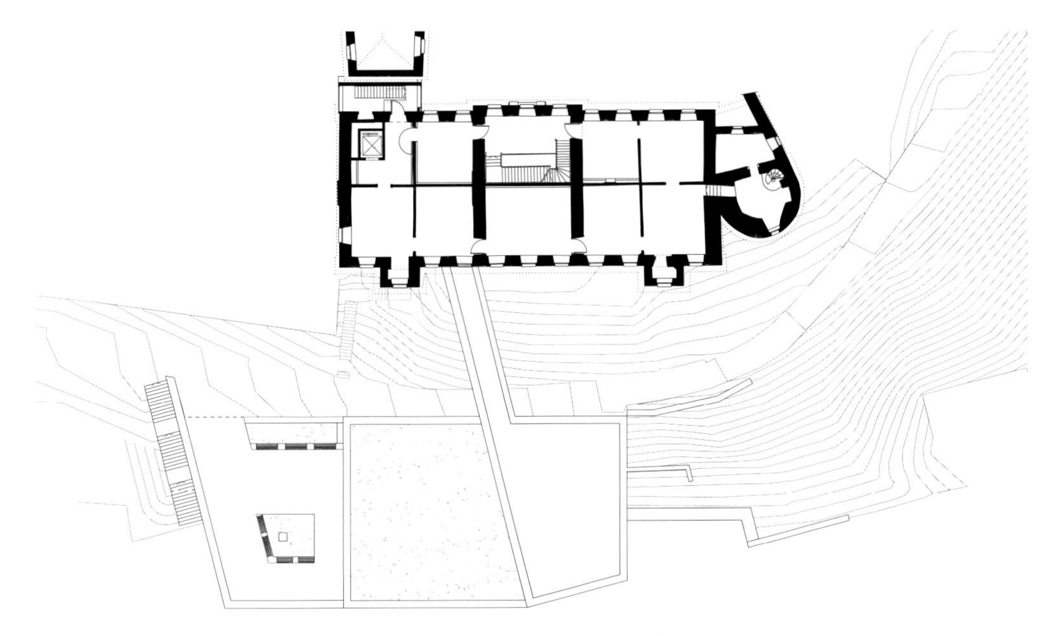

老建筑二层 first floor _ old building

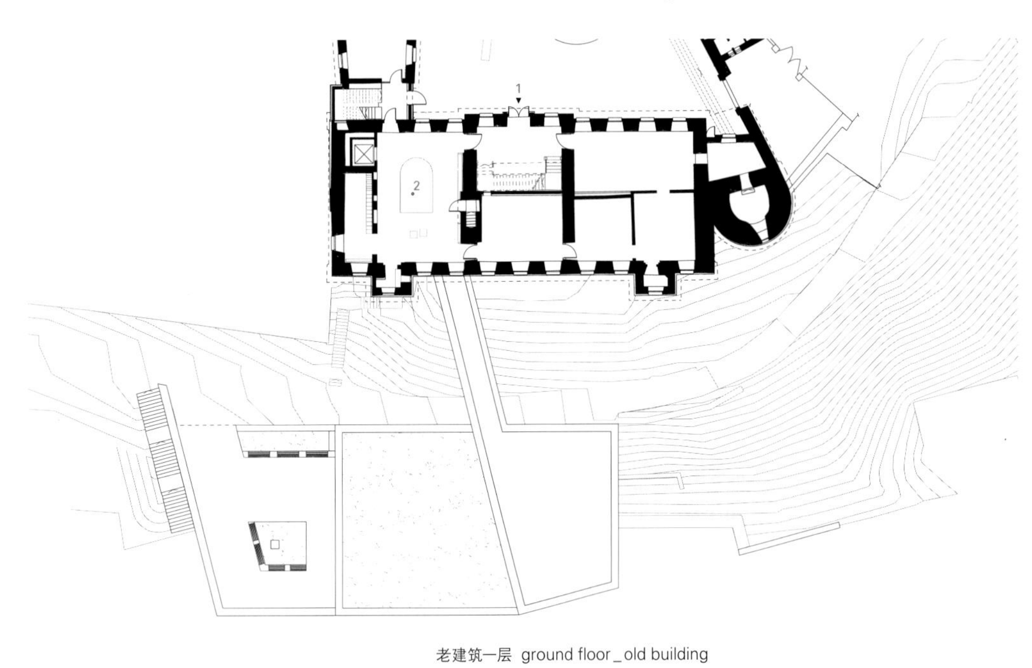

老建筑一层 ground floor _ old building

1. 主入口
2. 信息处/商店
3. 地下室
4. 通往新建筑的通道
5. 楼梯厅
6. 临时展区
7. 多功能厅/临时展区
8. 收发室
9. 卫生间
10. 储藏室

1. main entrance
2. information / shop
3. vault cellar
4. passage to the new building
5. stair hall
6. temporary exhibition
7. multipurpose hall / temporary exhibition
8. delivery
9. restrooms
10. storage

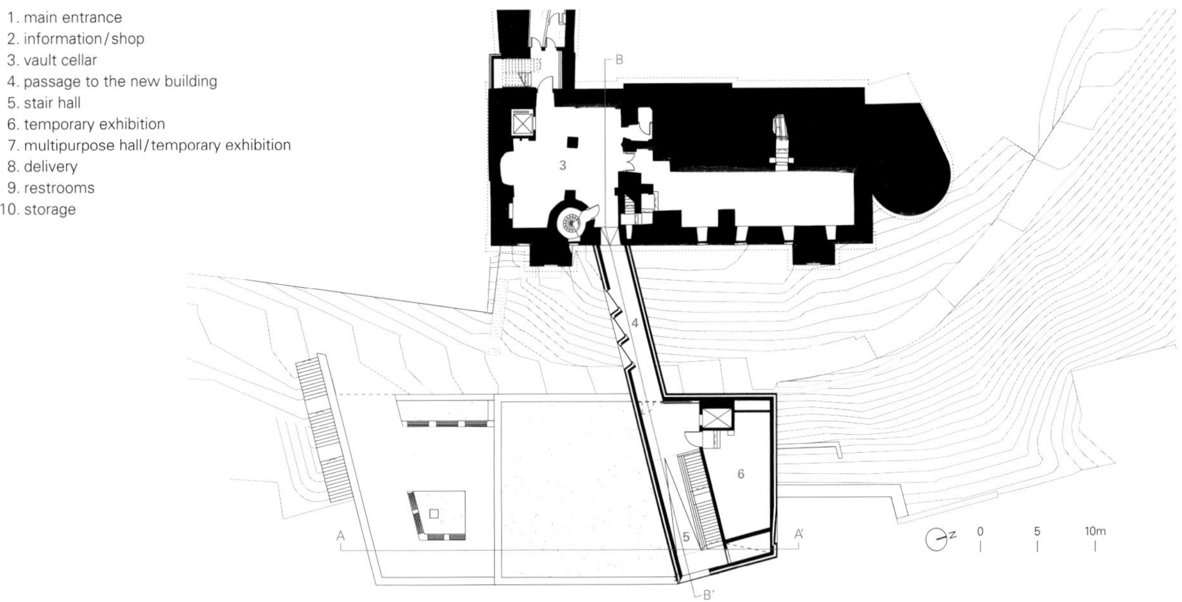

地下一层 first floor below ground

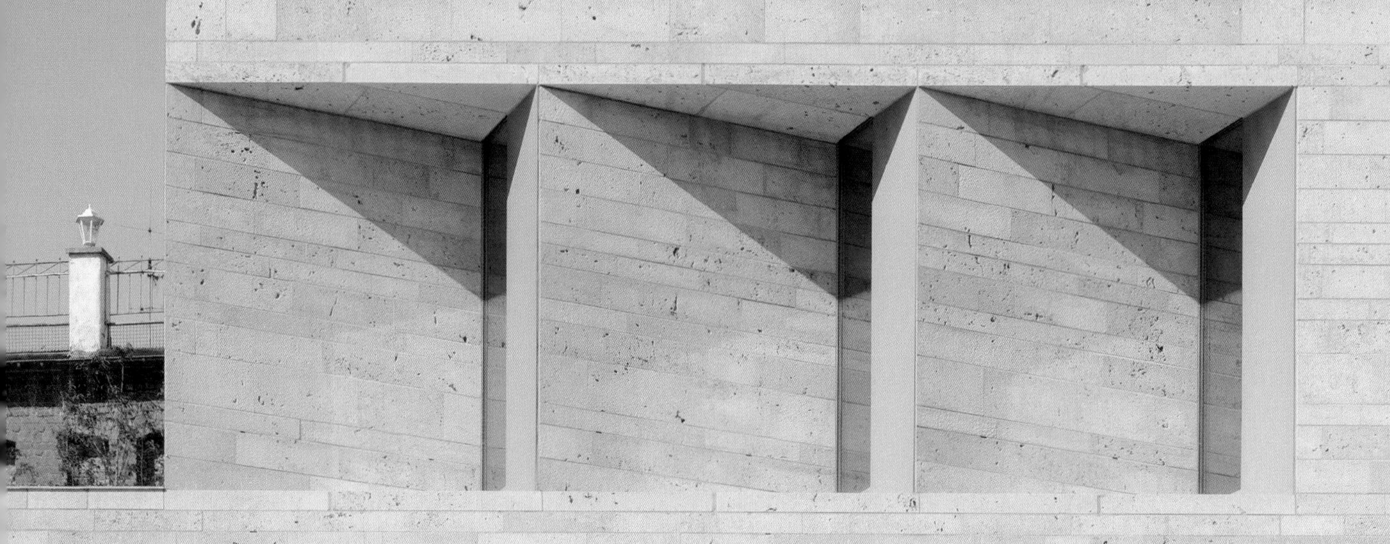
D-MUSEUM

EIGENE BILDWELTEN

The existing building connects to the extension by means of a bridge-like docking structure from the first basement of Landsberger Hof, accentuating this passage with three diagonally cut window openings towards Brückenplatz. The path leads straight to a full height, panoramic window, which opens up – from a height of 15m – an impressive view of the city. The "English Promenade", an existing historical footpath on the hill below Landsberger Hof, was preserved, and now passes under the connecting bridge to a public panoramic terrace on the lower roof of the museum.

The striking, staggered new building also serves as a mediating component in the city between the "Old Market" in the west and the significantly lower Ruhrstrasse to the east. By adopting the two main angles of Landsberger Hof and Ruhrstrasse, the new construction blends in naturally with its built environment. At the same time, Landsberger Hof – as a historically important palace on the city wall – remains dominant in the silhouette of the old town of Arnsberg. The homogeneously clad facade, with travertine from Gauingen in southern Germany, emphasizes the sculpturality of the new museum. Carefully incised three-dimensional window openings create an exciting relationship from the interior to the exterior space. For the design of the Museum and Cultural Forum only natural materials were used. The natural stone (travertine) for the monolithic facade and the oak floorboards for the exhibition spaces are sustainable, handcrafted materials of a high durability. The massive construction and the few window openings ensure a high thermal inertia, which is insensitive to temperature fluctuations and meets the requirements of a museum building very well.

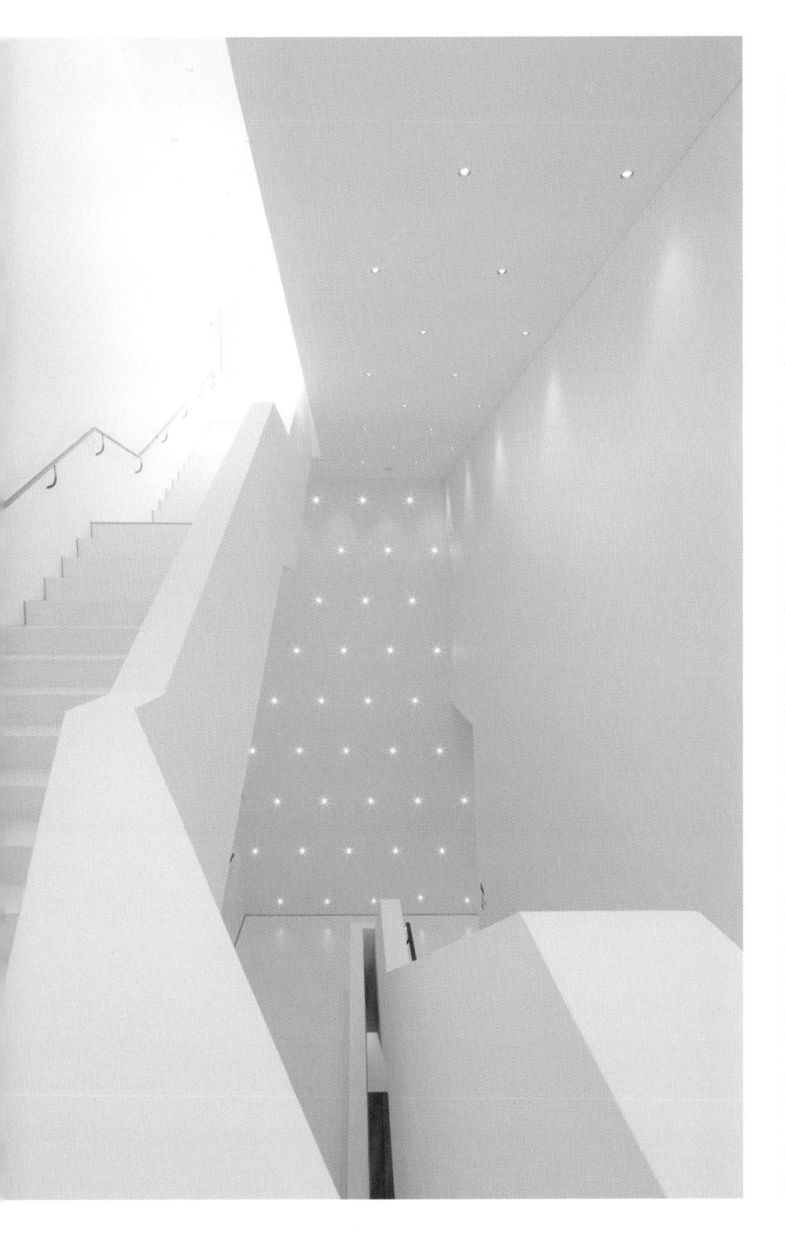

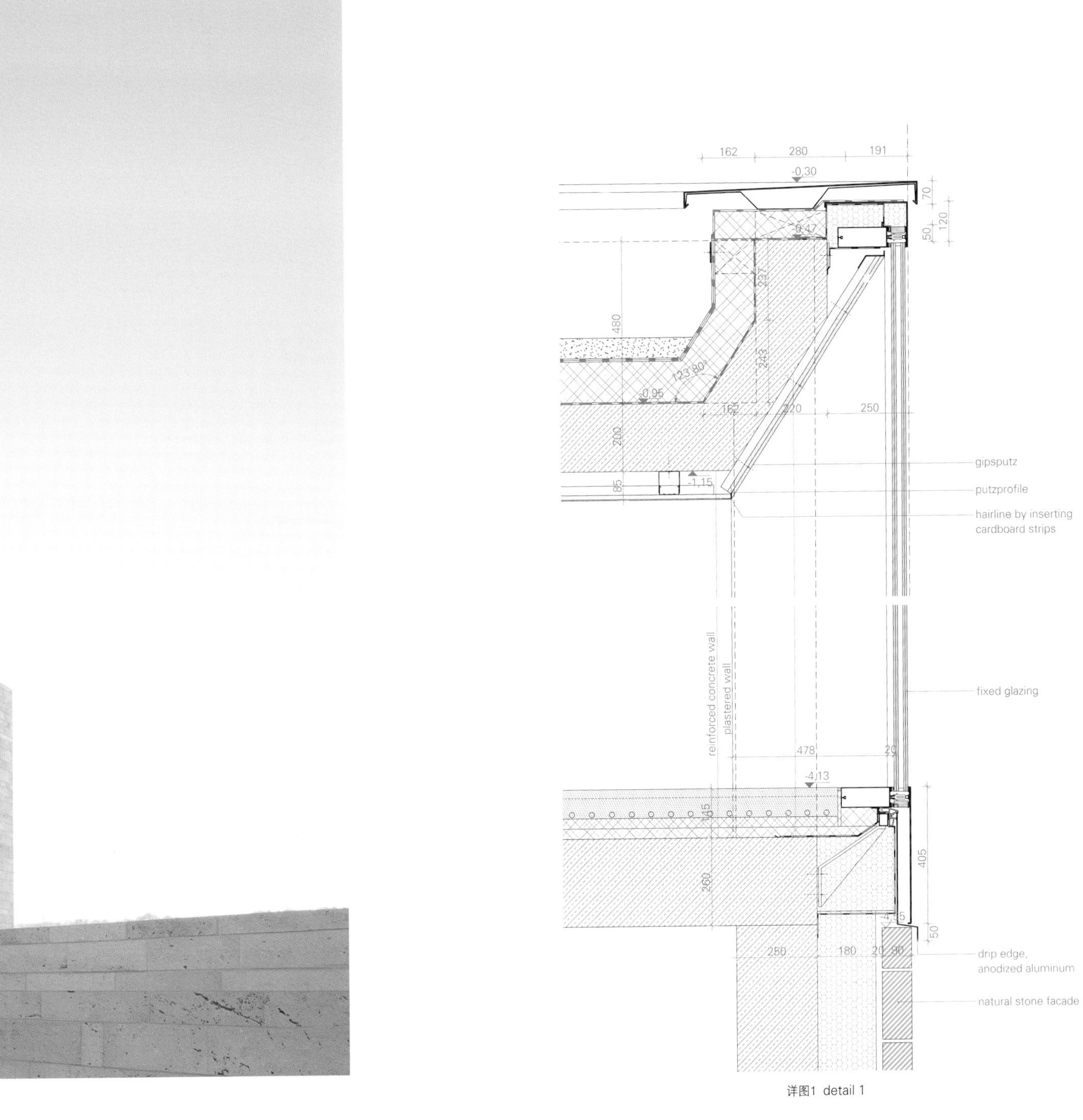

详图1 detail 1

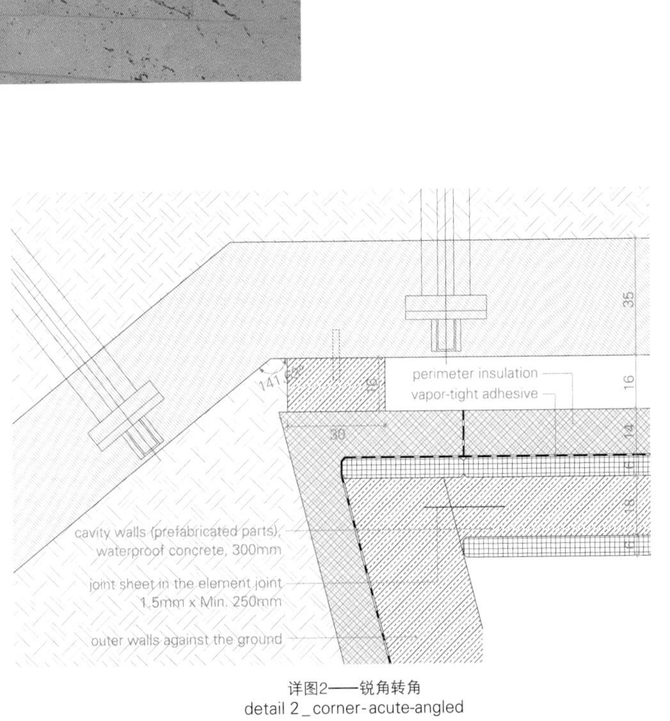

详图2——锐角转角
detail 2_corner-acute-angled

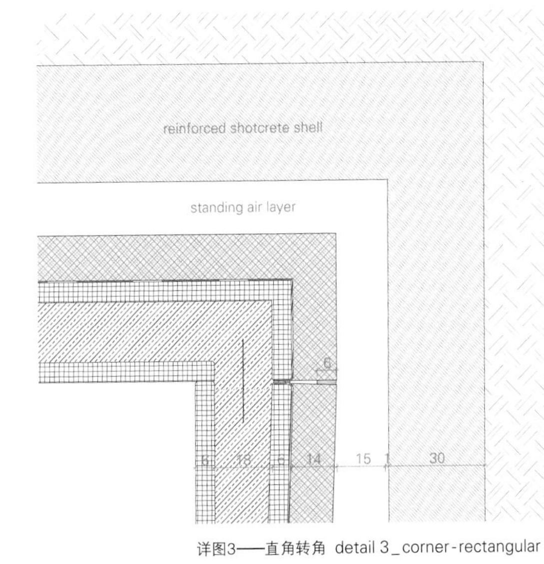

详图3——直角转角 detail 3_corner-rectangular

Das Sauerland, ein Burgenland
A Land of Castles

多边形画廊
Polygon Gallery

Patkau Architects

凸显北温哥华地方特色的新文化节点

A new cultural node reinforcing local identity of North Vancouver

四十多年来，"展示厅画廊"一直是北温哥华一个充满激情的独立摄影和媒体机构，而今，"多边形画廊"代表了它的重生。该建筑位于城市滨水新建区的前方，这里的基础设施被重新设计，文化也超越了过去的工业时代。

建筑物的主体部分拔地而起，为新的公共广场提供了一个开放式的通道，也带来了开阔的视野，可以远眺横跨伯拉德湾的温哥华天际线。其标志性的锯齿形轮廓是一个非凡而大胆的形式，上层为网眼铝质盖板，下层为几层反光不锈钢板。两种材料之间的相互作用赋予了这个单个体块一种短暂的深度，它能随着阳光的季节性变化和夜晚的氛围转换。

画廊负责人里德·希尔要求画廊空间没有固定障碍物，地板和墙壁可以分割，天花板上可以随意悬挂，随处都有电源和媒体，使用自然采光或可控电源。因此，画廊主体与其说是博物馆，不如说是一个工作室，有现成的工具随时用于创作。

建筑的结构格局起到了提升画廊高度和清晰空间感的双重作用，完全采用了从上方矩形屋顶漫射下来的北方光线进行采光，或明或暗。一组钢檩条系统可以提供照明、数据线、媒体装置、挂饰和临时隔断的轨道。连续的中央通道由坚固且易于修补的橡木地板铺成，便于通风和电气装置、数据线的安装，访客可以随时访问独立的作品和任意临时分区的配置。画廊的内部空间没有障碍，可以将它想象成一个现成的工具，能够容纳任何形式的艺术装置。

建筑上层包含一个大型灵活的宴会活动画廊，用于教育及外展服务。整个南墙是一面可开关的全景玻璃幕墙，可俯瞰伯拉德湾入海口和温哥华城。除了全玻璃的入口和大堂，下层还提供小型零售空间，辅助滨水区的发展。建筑的多功能应用使其成为滨水区日益增长的社交生活的焦点，并分享了画廊的能量。画廊给广场带来了活力，广场为北温哥华提供了一个全新的文化节点，凸显了毗邻大城市的小城市的地方特色。

The Polygon Gallery represents a rebirth of the Presentation House Gallery, which has been a passionately independent photography and media institution in North Vancouver for more than forty years. More site-maker than site response, the building stands at the front of an urban waterfront renewal site, where infrastructure is reimagined and culture outgrows an industrial past.

The main mass of the building is lifted from the ground plane to provide open access to both a new public plaza and a wide view of the Vancouver skyline across the Burrard Inlet. Its iconic saw-toothed profile, a singular and bold form, is clad in layers of mirrored stainless steel beneath expanded aluminum decking. The interplay between the two materials gives the

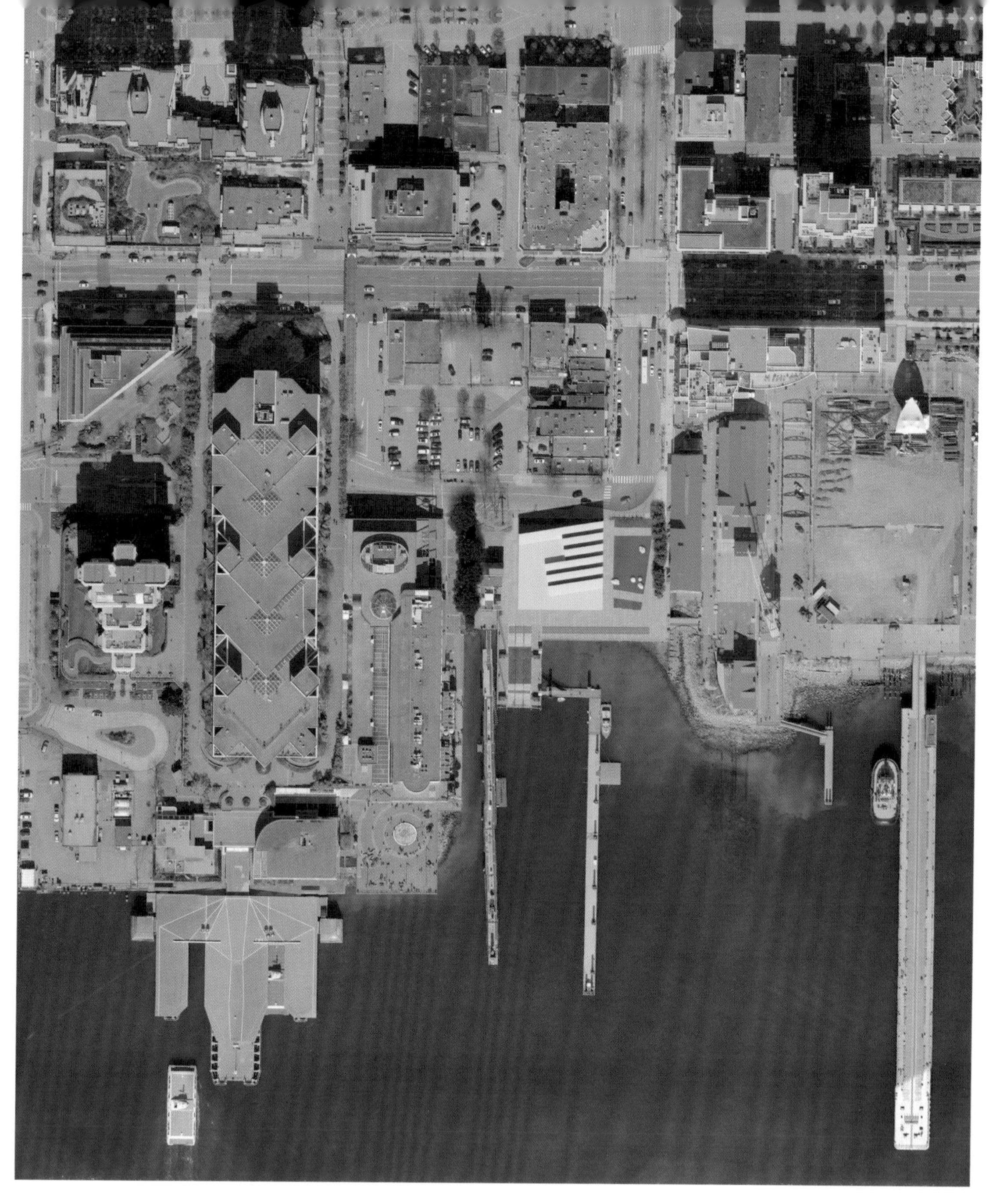

PACER

singular mass an ephemeral depth that shifts with seasonal sunlight and evening atmosphere.

Gallery Director Reid Shier requested the gallery space free of obstacles, with floors and walls that can be cut into, ceilings from which anything could be hung in any position, access to power and media anywhere, and lighting that can be natural or controlled. The main gallery is thus conceived as a ready instrument for creativity, more studio than museum. The structural musculature of the building performs the dual purposes of lifting the gallery and providing a clear space, completely daylit from above with diffuse northern light, or darkened. A system of steel purlins provides track for lighting, data, media, suspended works, and temporary partitions. The robust and easily patched oak flooring features a continuous central channel for ventilation, electrical, and data chases that give ready access to freestanding works and temporary partitions of any configuration. Within, the gallery space is conceived as a ready instrument, free of obstacles, and able to accommodate any form of art installation.

The upper level also contains a large flexible event gallery for education, outreach. Its entire southern wall is an operable glazed panorama overlooking Burrard Inlet and the city of Vancouver. In addition to the fully glazed entrance and lobby, the lower level supports small retail spaces, to help the waterfront development. These fine-grained uses make the building an attractor for a growing social life on the city's waterfront and share the energy of the gallery. The plaza, so activated by the gallery, provides a new cultural node for North Vancouver, reinforcing local identity for a small city that neighbors a larger and more prominent one.

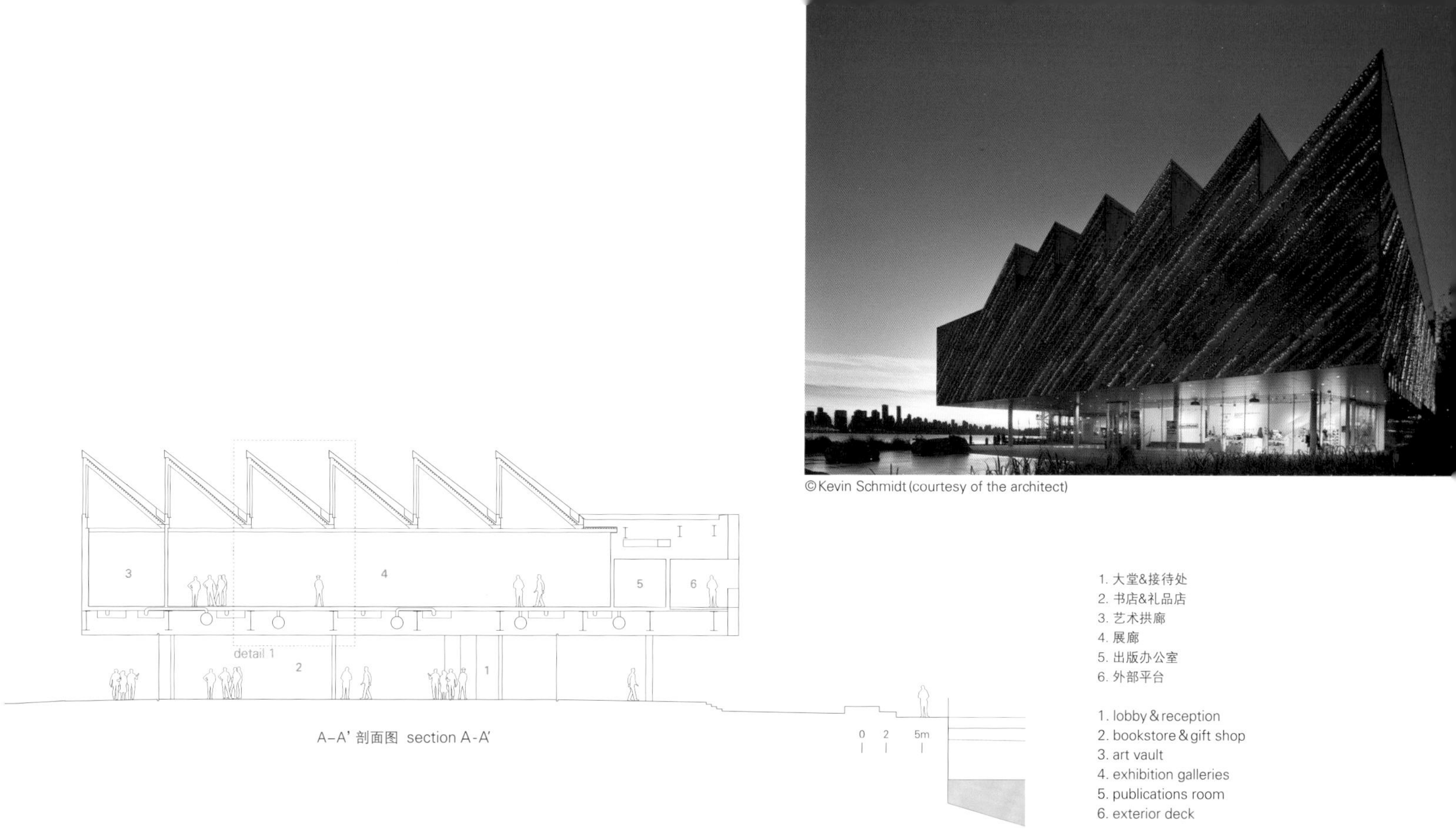

©Kevin Schmidt (courtesy of the architect)

A–A' 剖面图 section A-A'

1. 大堂&接待处
2. 书店&礼品店
3. 艺术拱廊
4. 展廊
5. 出版办公室
6. 外部平台

1. lobby & reception
2. bookstore & gift shop
3. art vault
4. exhibition galleries
5. publications room
6. exterior deck

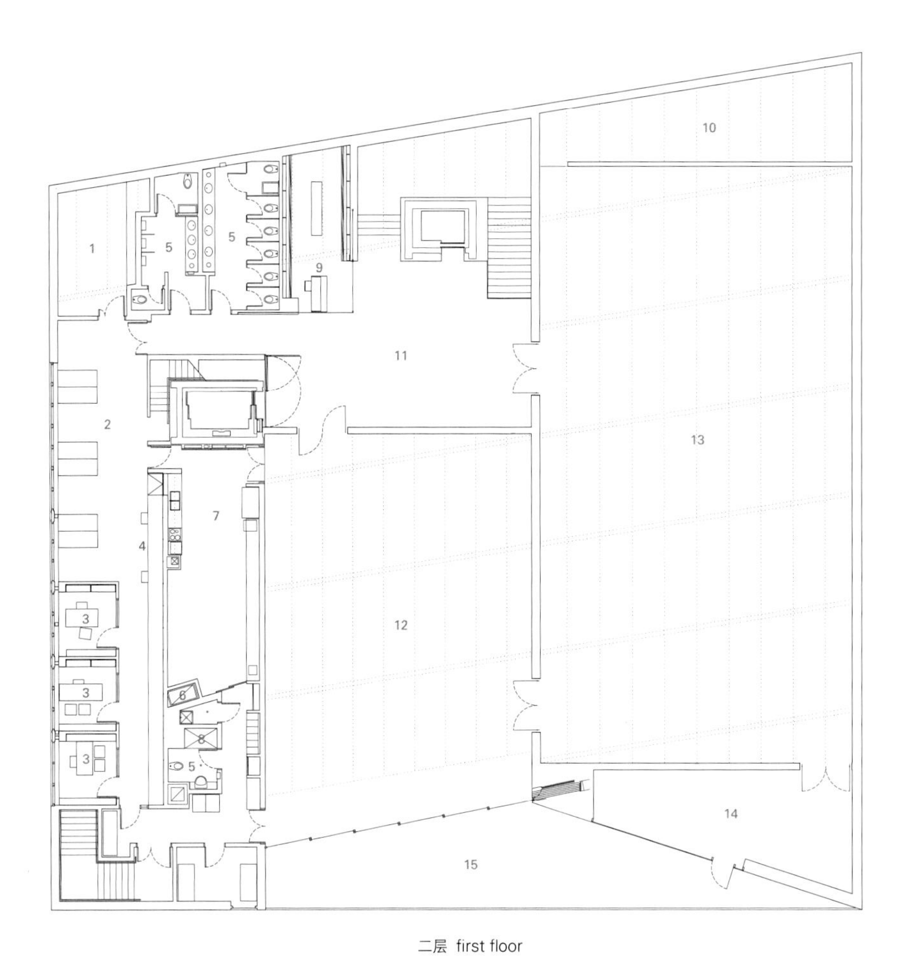

1. 准备室
2. 开放式办公室
3. 办公室
4. 公用办公桌
5. 洗手间
6. 活动储藏室
7. 午餐及准备室
8. 淋浴间
9. 书店
10. 艺术拱廊
11. 大堂画廊
12. 活动画廊
13. 展廊
14. 出版办公室
15. 外部平台

1. framing & preparation
2. open office
3. office
4. hot desks
5. washroom
6. event storage
7. lunch & prep
8. shower
9. bookstore
10. art vault
11. foyer gallery
12. event gallery
13. exhibition galleries
14. publications room
15. exterior deck

二层 first floor

1. 大堂&接待处
2. 书店&礼品店
3. 走廊&更衣室
4. 商业零售单元
5. 工作坊&储藏室
6. 储藏室&装卸处
7. 回收处
8. 电气室
9. 通信室
10. 机械设备间
11. 自行车安全存储
12. 卫生间
13. 货梯
14. 客梯
15. 未来CRU

1. lobby & reception
2. bookstore & gift shop
3. corridor & coat/lockers
4. commercial retail unit
5. workshop & storage
6. storage & loading
7. recycling
8. electrical room
9. communications room
10. mechanical room
11. secure bike storage
12. WC
13. freight elevator
14. elevator
15. future CRU

0 2 5m

一层 ground floor

项目名称：Polygon Gallery / 地点：North Vancouver, British Columbia, Canada / 建筑师：Patkau Architects Inc.
项目团队：John Patkau, Patricia Patkau, Peter Suter, Mike Green, Jackie Ho, Marc Holland, Haley Zhou / 结构：Fast&Epp Structural Engineers
机电：Integral Group / 声学：Dan Lyzun and Associates / 代码顾问：LMDG / 造价顾问：Turnbull Construction Services
客户：Presentation House Gallery / 面积：2,100m² / 设计与竣工时间：2013—2016 / 摄影师：©James Dow (courtesy of the architect), except as noted

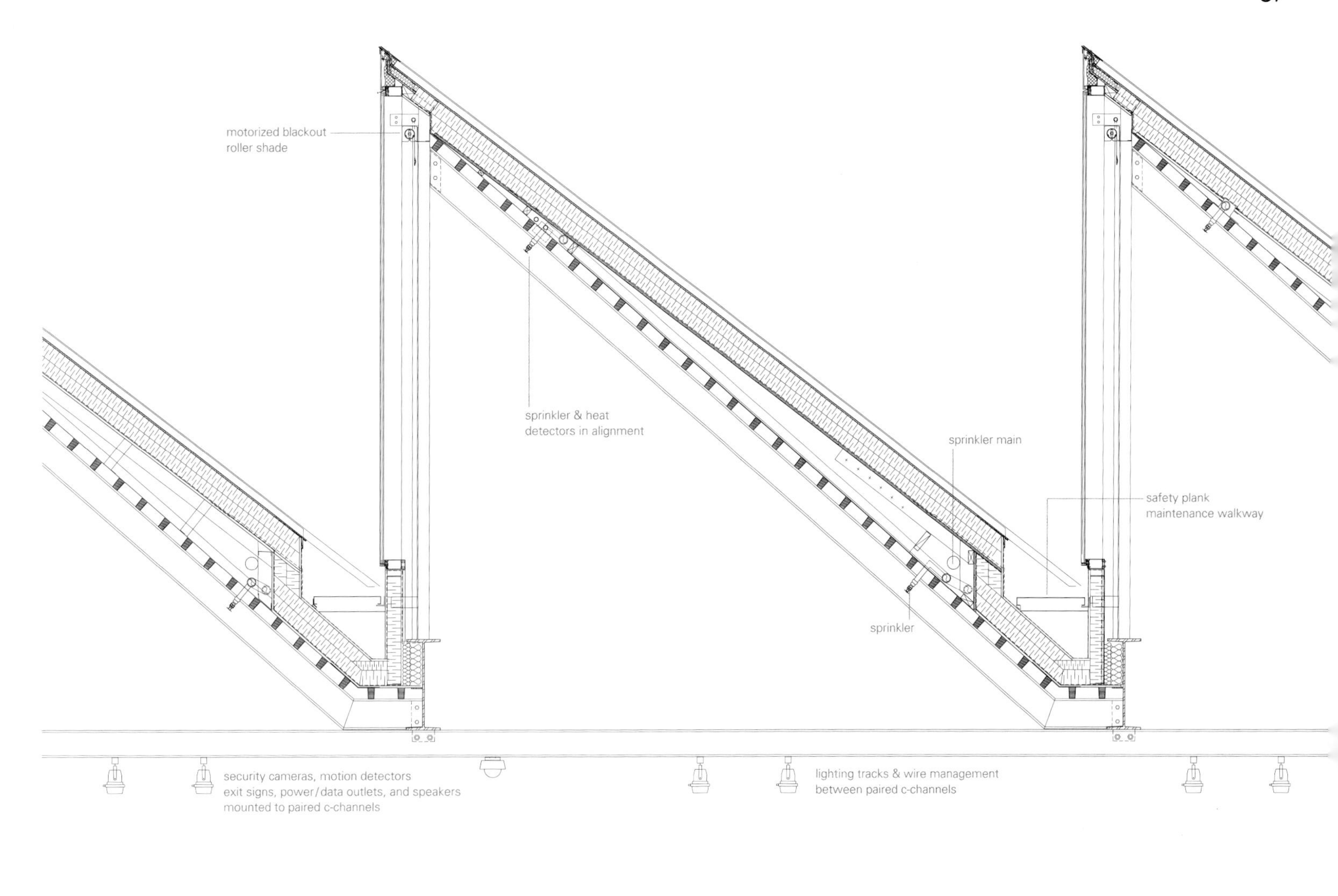

typical roof assembly
- concealed fastener metal roofing panel
- composite drainage layer
- self-adhered vapor permeable waterproof membrane underlay
- 180mm Polyiso insulation
- vapor barrier membrane
- 15mm treated plywood sheathing
- 38mm wide treated tapered furring
- acoustic insulation infill strips (in deck flutes)
- acoustic steel roof deck
- exposed steel structure

typical floor assembly
- engineered oak hardwood flooring
- 203mm deep steel deck with 65mm concrete topping over steel floor beams C/W spray-applied fireproofing (beams only)
- suspended perforated sheet aluminum ceiling panels

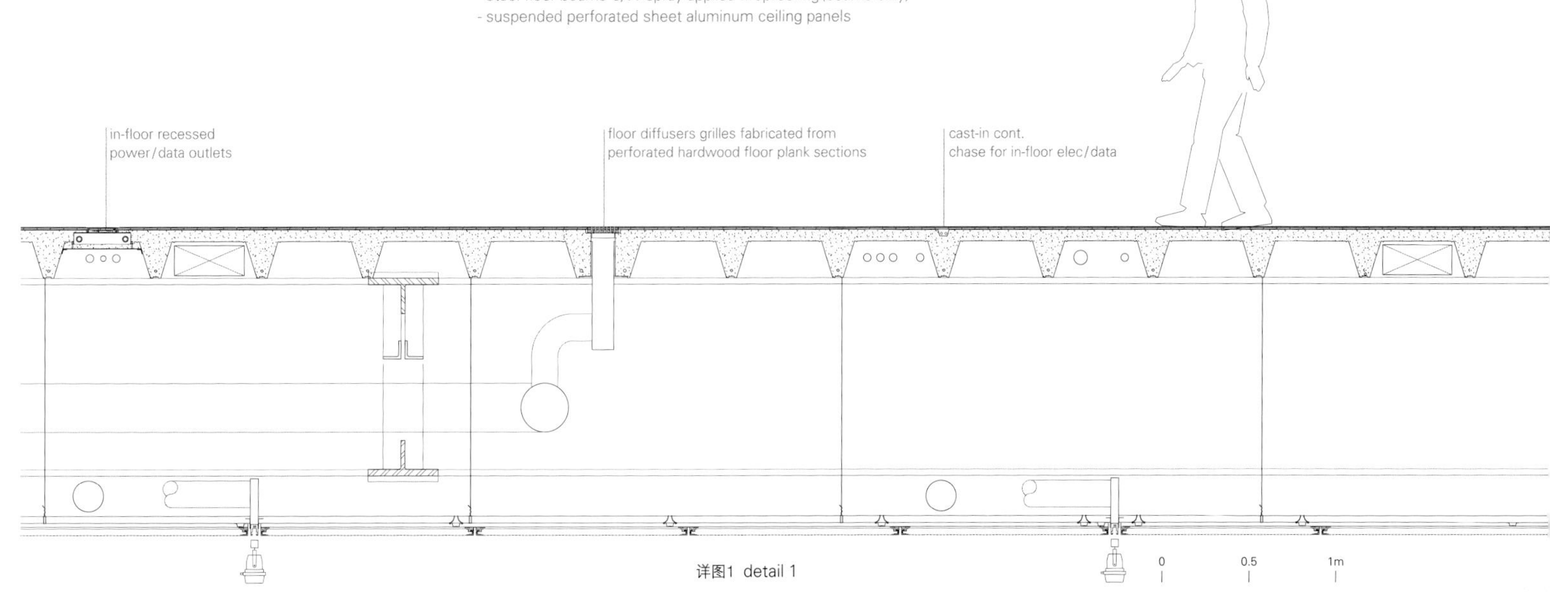

详图1 detail 1

扭体博物馆
The Twist

BIG

河对岸的扭体博物馆构建了景观中的文化之路
Twisted gallery across the river completes a cultural path in landscape

扭体博物馆（The Twist）是BIG在挪威的第一个项目，它作为一座可以栖居的桥梁，在中部以扭转的身姿跨越蜿蜒的Randselva河，为Jevnaker的Kistefos雕塑公园带来全新的艺术体验。作为公园内最新落成的、面积为1000m²的当代艺术机构及设施，扭体博物馆连接了两个草木丛生的河岸，完善了贯穿北欧最大雕塑公园的文化路线。

扭体博物馆坐落在一家历史悠久的纸浆厂附近，被构思为一个在中间部位扭转90°的"横梁"，在Randselva河的上方创造出一种雕塑般的形式。公园中的游客一边漫步，一边欣赏由Anish Kapoor、Olafur Eliasson、Lynda Benglis、Yayoi Kusama、Jeppe Hein和Fernando Botero等国际艺术家专门为场地设计的作品，最后将穿过扭体博物馆完成整个艺术之旅。作为公园的第二座桥梁和自然环境的延伸，新的博物馆在改变游客体验的同时，也使得Kistefos雕塑公园的室内展览空间扩增了一倍。

建筑体量通过简单的扭转，得以从地势较低的森林河岸延伸至北面地势较高的山坡地带。作为景观中的连续路径，建筑的两侧都可以充当主入口。游客可从南入口穿过一座16m长的铝面钢桥到达一个双层高的空间，从这里可以清晰地看到北端的风景。同时还有一个9m长的人行天桥与该空间相连。博物馆的双曲线几何形式由40cm宽的铝板像书籍一样排列而成，并以扇动的姿势轻柔地移动和变化。建筑内部也采用了同样的原理，地面、墙壁和天花板均覆盖以8cm宽的涂白的杉木板条，为Kistefos举办的一系列挪威国内和国际短期展览提供了统一的背景。游客可以从任何一个方向来体验这座扭转的画廊，就像是穿过照相机快门一样。

建筑北端的通高玻璃墙为博物馆带来了看向纸浆厂与河流的全景视野，同时以扭转的姿态上升，形成一个25cm宽的带状天窗。玻璃窗的弯曲形式为博物馆内部带来了多种多样的光照氛围，并因此形成了三个与众不同的画廊：其一是以自然光照亮的位于北面的画廊，拥有宽敞的空间和全景视野；其二是天花板较高、光线昏暗、采用人工照明的南侧画廊；其三是位于前两者之间的雕塑般的空间，拥有一个扭曲的长条形屋顶天窗。划分、分割或合并画廊空间的能力，为Kistefos雕塑公园的艺术项目创造了灵活性。玻璃楼梯连接了位于北部河岸的地下楼层，建筑的铝质底面在这里成了地下室和卫生间区域的天花板。全宽的玻璃立面拉近了游客与下方的河流之间的距离，为奥斯陆郊外林地的悠闲之旅增添了沉浸式的体验。

Traversing the winding Randselva river, BIG's first project in Norway – The Twist – opens as an inhabitable bridge torqued at its center. It forms a new journey and art piece within the Kistefos Sculpture Park in Jevnaker, Norway. Kistefos' new 1,000m² contemporary art institution doubles as infrastructure to connect two forested riverbanks, completing the cultural route through northern Europe's largest sculpture park.
Built around a historical pulp mill, The Twist is conceived as a beam warped 90 degrees near the middle to create a sculptural form as it spans the Randselva. Visitors roaming the park's site-specific works – by international artists such as Anish Ka-

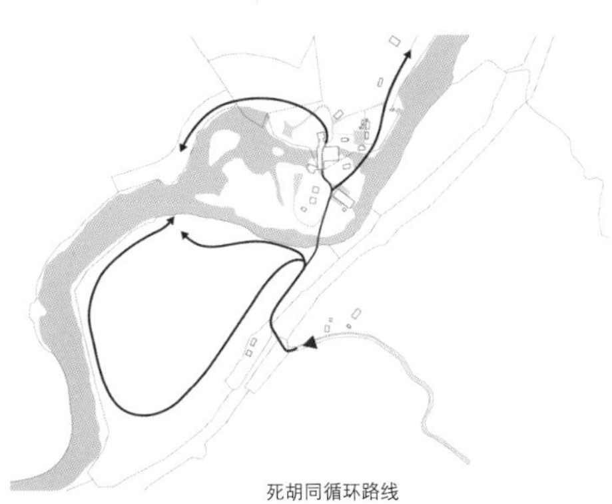

死胡同循环路线

DEAD END CIRCULATION

The river is the natural divide between the two sides of the Kistefos sculpture park, creating circulation issues.

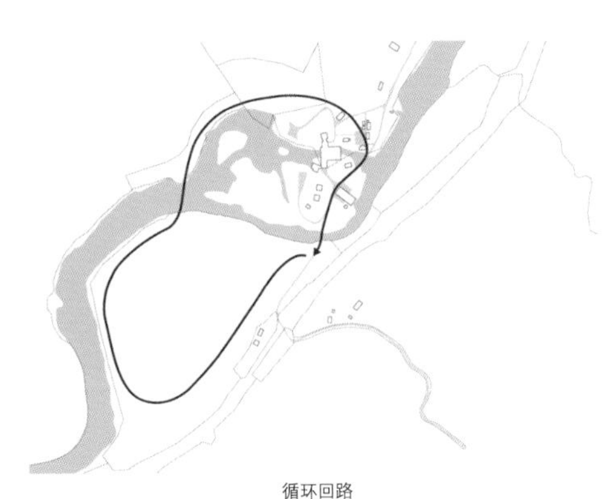

循环回路

ONE CIRCULATION LOOP

We suggest adding a new bridge to tie the area together, connecting the landscape and interior galleries in a natural continuous loop.

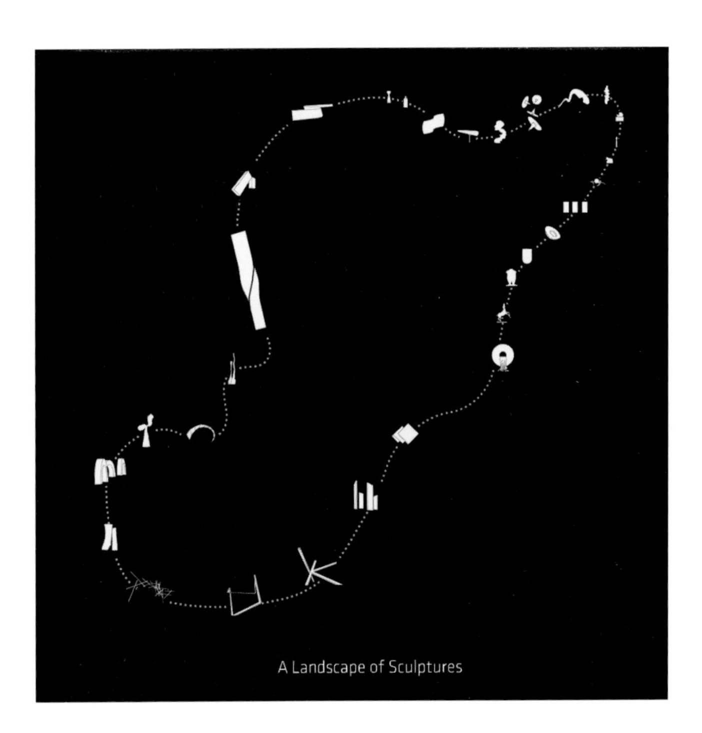

建筑/博物馆
Building / Museum

基础设施/桥梁
Infrastructure / Bridge

艺术/雕塑
Art / Sculpture

艺术桥梁
Art Bridge

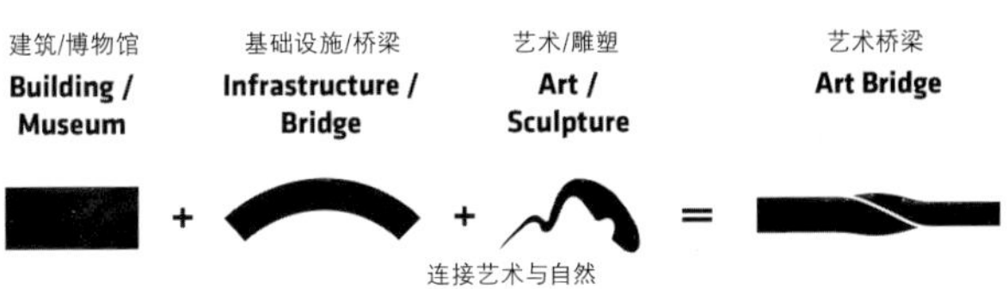

连接艺术与自然

BRIDGING ART AND NATURE

We propose a hybrid of architecture, infrastructure and sculpture.

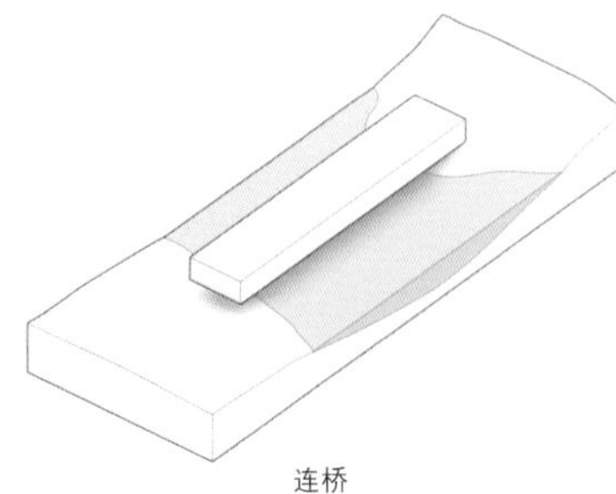

连桥

BRIDGING

The total required building volume is placed as a bridge spanning the Randselva river, connecting the two edges of the site.

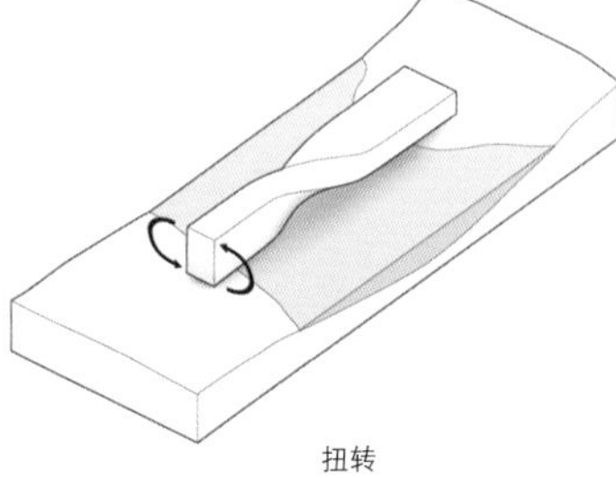

扭转

TWISTING

A simple twist in the building volume allows the bridge to lift from the relatively lower forested area towards the south up to the hillside area in the north.

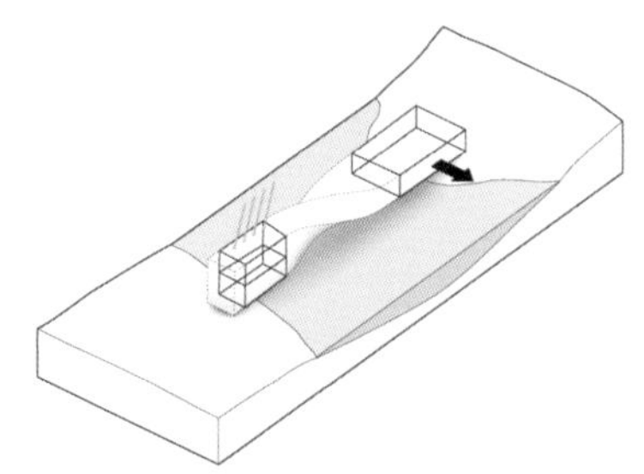

堆叠的垂直&水平画廊

STACKED VERTICAL & HORIZONTAL GALLERIES

As a result of the building's twist, vertical galleries in the south enjoy natural light from overhead while the large horizontal, open gallery to the north offers views of the historic pulp mill and river.

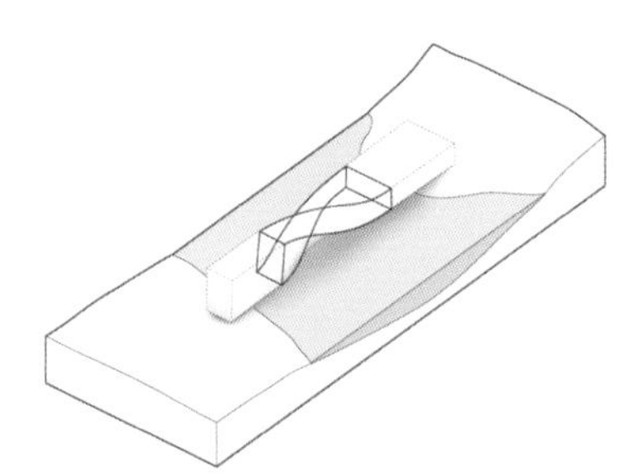

反映文脉关系的画廊

CONTEXTUAL GALLERIES

The twisted geometry in the middle of the building merges the vertical and horizontal in a single motion, reflecting the landscape.

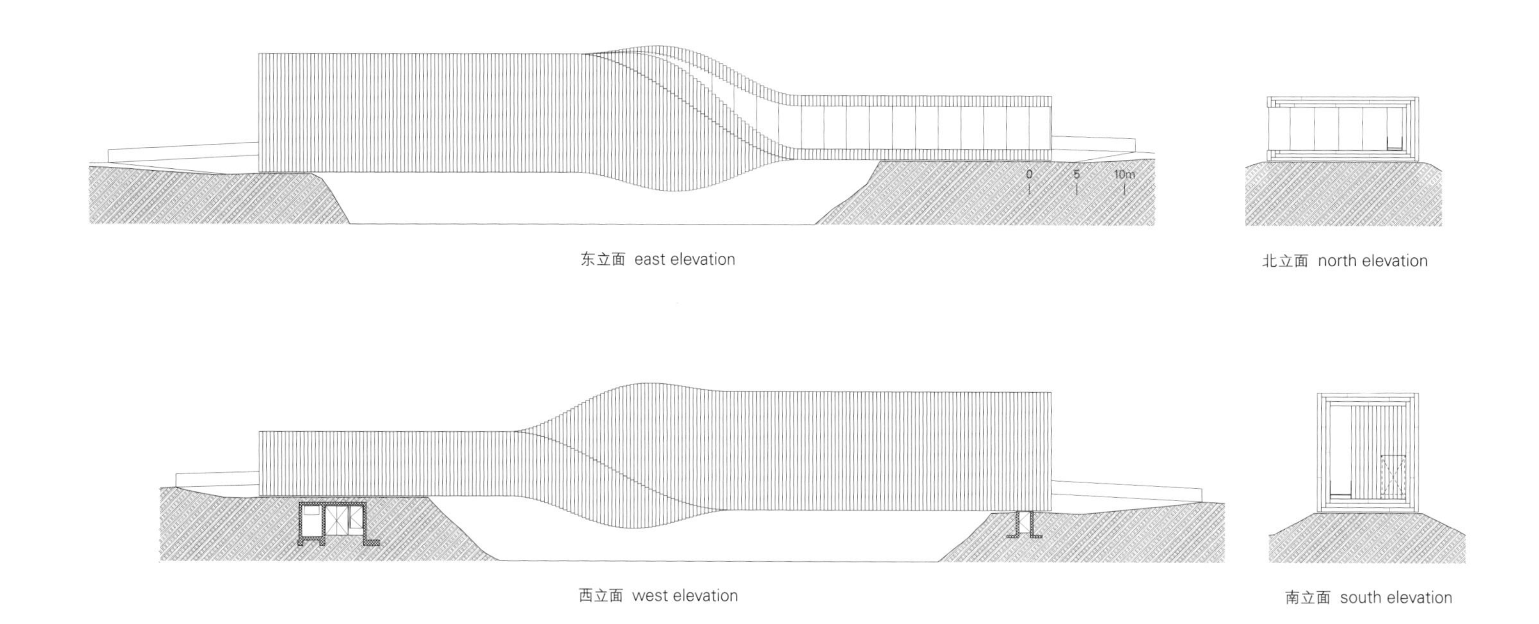

东立面 east elevation

北立面 north elevation

西立面 west elevation

南立面 south elevation

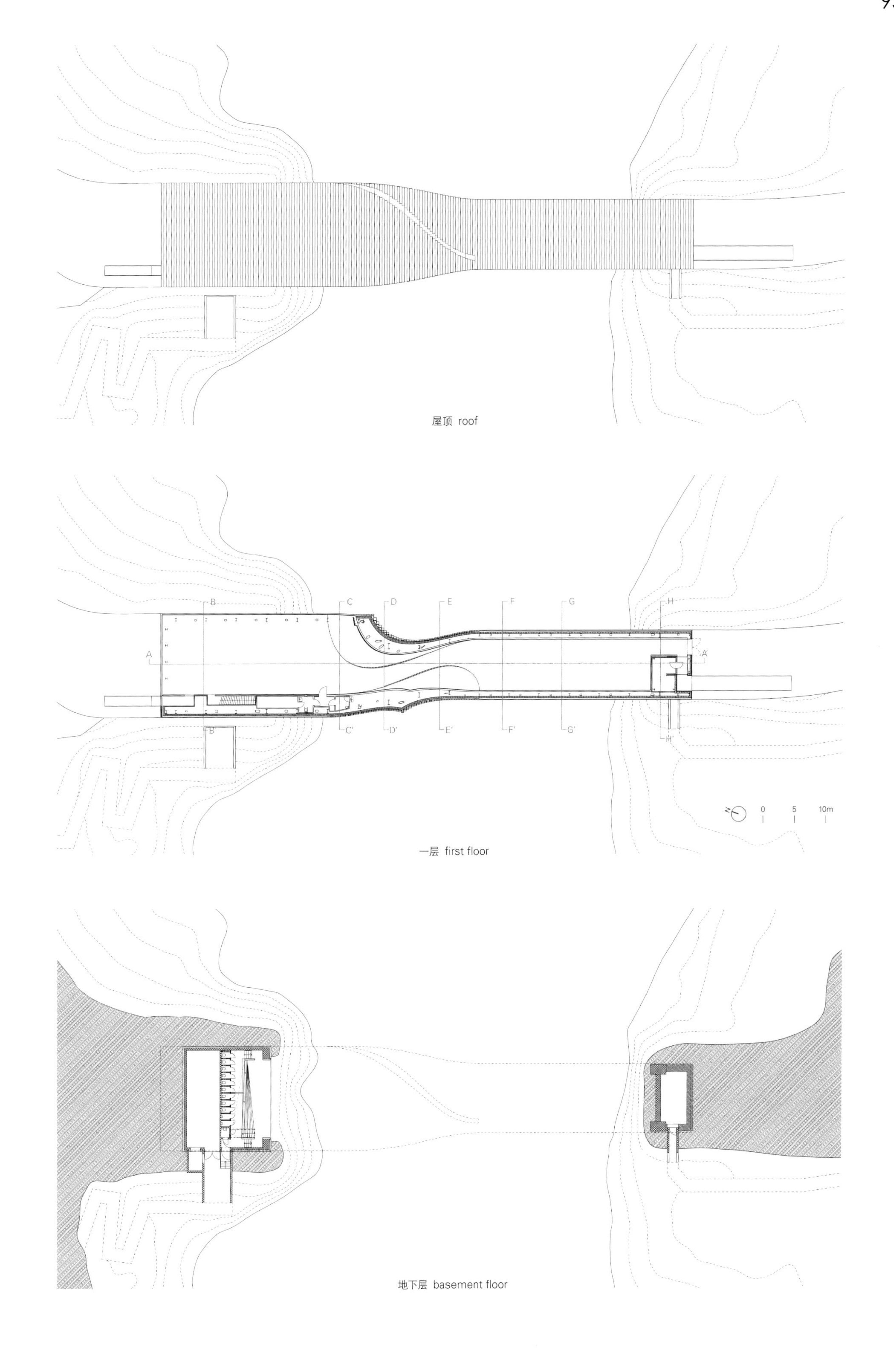

屋顶 roof

一层 first floor

地下层 basement floor

poor, Olafur Eliasson, Lynda Benglis, Yayoi Kusama, Jeppe Hein and Fernando Botero, among others – cross The Twist to complete the art tour. As a second bridge and natural extension to the park, the new museum transforms the visitor experience while doubling Kistefos' indoor exhibition space.

A simple twist in the building's volume allows the bridge to lift from the lower, forested riverbank in the south, up to the hillside area in the north. As a continuous path in the landscape, both sides of the building serve as a main entrance. From the south entry, visitors cross a 16m aluminum-clad steel bridge to reach the double-height space, with a clear view to the north end, similarly linked with a 9m pedestrian bridge. The double-curve geometry of the museum is comprised of straight 40cm wide aluminum panels arranged like a stack of books, shifted ever so slightly in a fanning motion. The same principle is used inside, with 8cm wide fir slats, painted white, cladding the floor, wall and ceiling to create one uniform backdrop for Kistefos' short-term Norwegian and international exhibitions. From either direction, visitors experience the twisted gallery as though walking through a camera shutter.

On the north end, a full-height glass wall, offering panoramic views to the pulp mill and river, tapers while curving upwards to form a 25cm wide strip of skylight. Due to the curved form of the glass windows, the variety of daylight entering the museum creates three distinctive galleries: a wide, naturally lit gallery with panoramic views on the north side; a tall, dark gallery with artificial lighting on the south side; and, in between, a sculptural space with a twisted sliver of roof light. The ability to compartmentalize, divide or merge the gallery spaces creates flexibility for Kistefos' artistic programming. A glass stairway leads down to the museum's lower level on the north river embankment, where the building's aluminum underside becomes the ceiling for the basement and restroom area. Another full-width glass wall brings visitors even closer to the river below, enhancing the overall immersive experience of being in the idyllic woodlands outside of Oslo.

项目名称：The Twist / 地点：Jevnaker, Norway / 建筑师：BIG
主管合伙人：Bjarke Ingels, David Zahle / 项目主管：Eva Seo-Andersen
项目建筑师：Mikkel Marcker Stubgaard / 项目团队：Aime Desert, Alberto Menegazzo, Aleksandra Domian, Aleksandra Sobczyk, Alessandro Zanini, Alina Tamosiunaite, Andre Zanolla, Balaj Alin Ilulian, Brage Mæhle Hult, Brian Yang, Carlos Ramos Tenorio, Carlos Surrinach, Casey Tucker, Cat Huang, Channam Lei, Christian Dahl, Christian Eugenius Kuczynski, Claus Rytter Bruun de Neergaard, Dag Præstegaard, David Tao, Edda Steingrimsdottir, Espen Vik, Finn Nørkjær, Frederik Lyng, Jakob Lange, Joanna M. Lesna, Kamilla Heskje, Katrine Juul, Kekoa Charlot, Kei Atsumi, Kristoffer Negendahl, Lasse Lyhne-Hansen, Lone Fenger Albrechtsen, Mads Mathias Pedersen, Mael Barbe, Marcelina Kolasinska, Martino Hutz, Matteo Dragone, Naysan John Foroudi, Nick Huizenga, Nobert Nadudvari, Ovidiu Munteanu, Rasmus Rosenblad, Richard Mui, Rihards Dzelme, Roberto Fabbri, Ryohei Koike, Sofia Rokmaniko, Sunwoong Choi, Tiina Liisa Juuti, Tomas Ramstrand, Tore Banke, Tyrone Cobcroft, Xin Chen
合作方：AKT II, ÅF Belysning, AS Byggeanalyse, Baumetall Design, BIG Ideas, Bladt Industries, Brekke & Strand, Davis Langdon, DIFK, ECT, Element Arkitekter, Erichsen & Horgen, Fokus Rådgivning, GCAM, Grindaker, Lüchinger & Meyer, Max Fordham, MIR, Rambøll / 客户：Kistefos Museum / 用途：culture / 建筑面积：1,000m^2
竣工时间：2019 / 摄影师：©Laurian Ghinitoiu (courtesy of the architect)

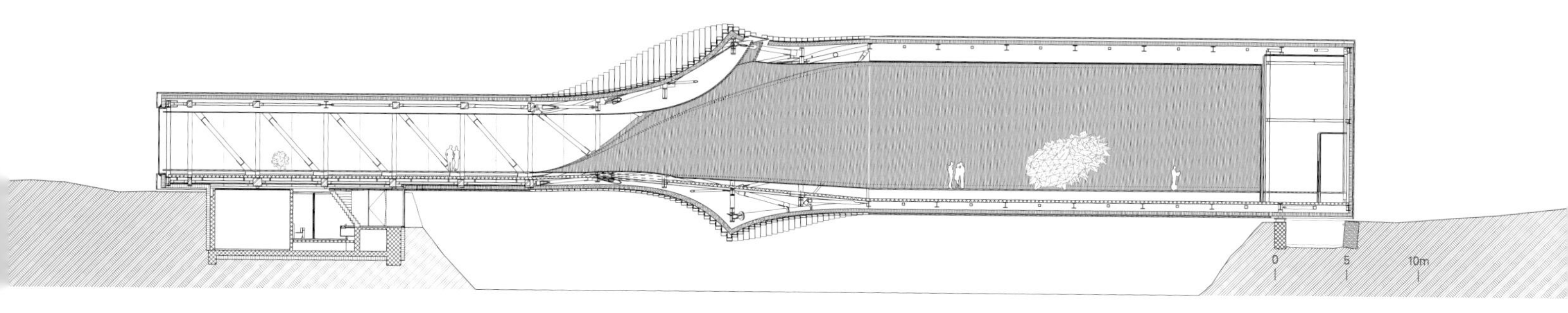

A–A' 剖面图 section A-A'

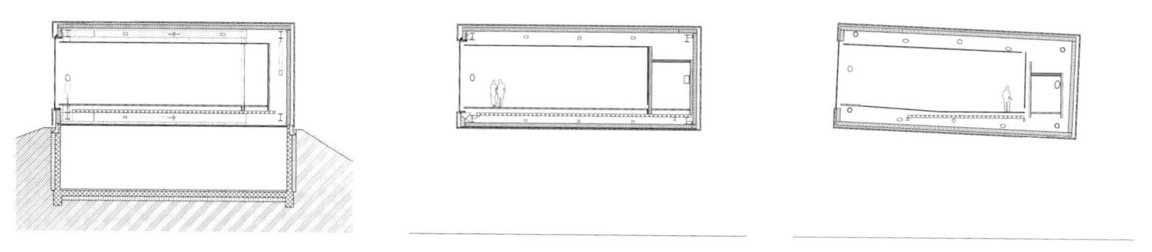

B–B' 剖面图 section B-B'

C–C' 剖面图 section C-C'

D–D' 剖面图 section D-D'

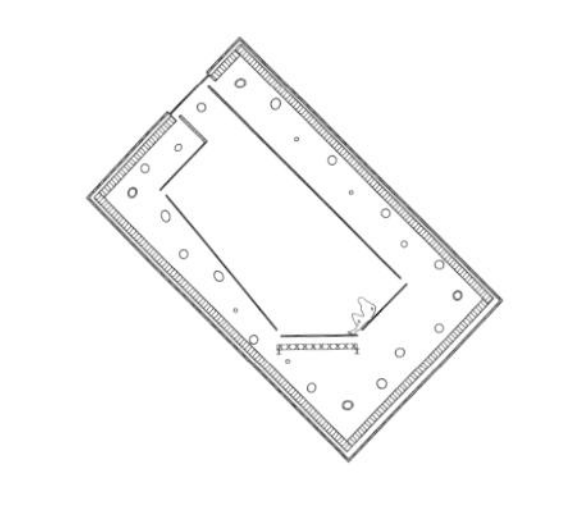

E-E' 剖面图 section E-E'

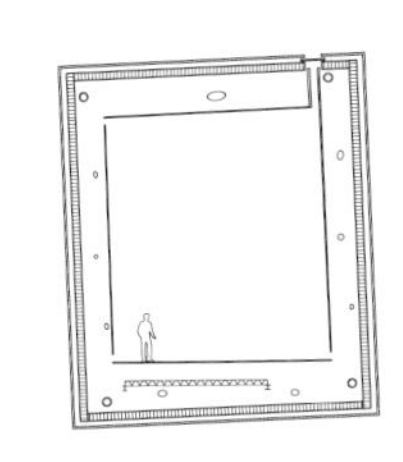

F-F' 剖面图 section F-F'

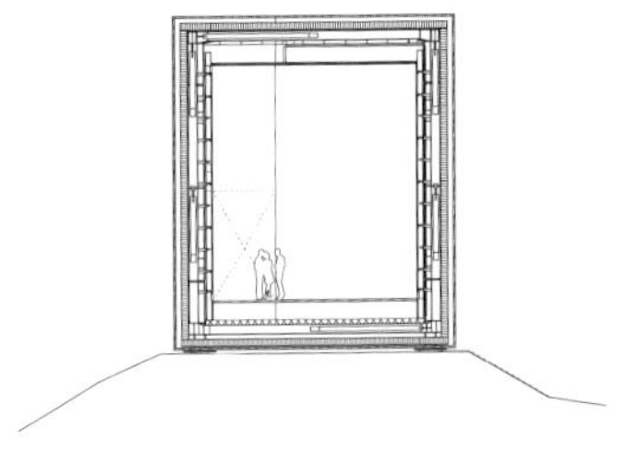

G-G' 剖面图 section G-G'

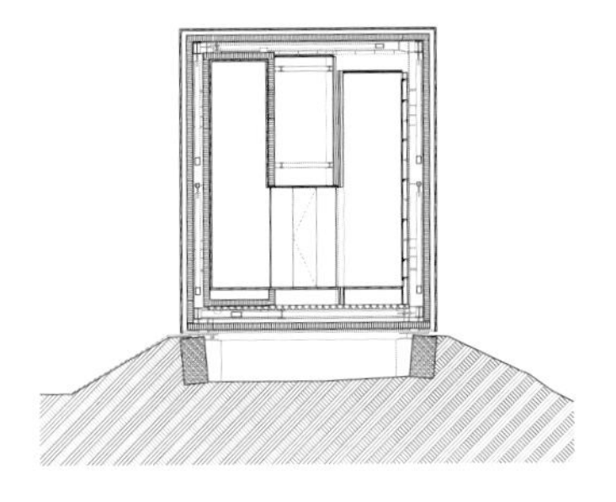

H-H' 剖面图 section H-H'

当学习遭遇教育综合体

Learning Complex

本文探索了来自墨西哥、秘鲁和越南的三个教育建筑实例，其设计者分别是：墨西哥的Ignacio Urquiza、Bernardo Quinzaños+Rodrigo Valenzuela Jerez、Camilo Moreno；越南的Vo Trong Nghia Architects建筑师事务所和秘鲁的Barclay&Crousse Architecture建筑事务所。2019年，这三个项目为这几家事务所赢得了国际认可，几家事务所在建筑特点上具有共性。一方面，从城市规划的角度来看，校园被设计成拥有道路、建筑和公共空间的小城市。这些空间

This article explores three examples of educational architecture from Mexico, Peru and Vietnam. These three architectural practices - Ignacio Urquiza, Bernardo Quinzaños + Rodrigo Valenzuela Jerez, Camilo Moreno (Mexico); Vo Trong Nghia Architects (Vietnam) and Barclay & Crousse Architecture (Peru) have achieved international recognition in 2019 for these projects, which share certain commonalities in their architectural designs. On the one hand, from an urban planning point of view, the campuses have been designed as small cities, with paths, buildings, and public spaces. They recognise the autonomy of these centers, spaces which emerged in the fringes of cities, which can create their own geometrical boundaries outside the organic growth of their

当学习遭遇教育综合体_Learning Encounter Complexes / Ana Souto

Encounter

出现在城市的边缘，可以在邻近城市的有机增长之外创建自己的地域界限，形成各自的自治中心。另一方面，这个建筑设计方法将促进人与人的自由碰撞与交流视为焦点，创建内部和户外空间的相互影响，如光与影的交互。同时，充分利用所使用的材料：清水混凝土、传统的砖块、植被，尤其是自然光。最后，所有项目都展示出一个清晰的环境理念，选择符合当地文化和气候的设计，提倡当代乡土建筑的可持续设计。

neighbouring cities. On the other hand, the architectural approach has focused on inviting and facilitating casual encounters and the exchange of ideas, by creating interplays of inside and outdoors spaces, light and shadow, as well as making the most of the materials used: playing with the bare concrete, traditional brick, vegetation, and especially, with light. Finally, all projects demonstrate a clear understanding of their contexts, and have chosen designs that embrace the culture and climate of their locations, promoting a contemporary vernacular architecture which promotes sustainable designs.

当学习遭遇教育综合体
Learning Encounter Complexes

Ana Souto

"当你规划一所学校时，你是否觉得你将拥有七个研讨室……还是说它是具有某种特质的地方，能让你身在其中受到启发？以某种方式在那里交谈，并获得一种交流感受？室内会有壁炉吗？应该有一个画廊，而不是走廊。画廊真的成了学生们的教室，在那里，总有听不懂老师讲课的男孩。然而，他可以和另一个男孩交流，那个男孩和他不同，能听懂老师的授课，这样一来，两人就都能理解授课内容了。"　——路易斯·康，《与学生的对话》（1969年）

在过去的两个世纪里，用于教学目的的建筑类型发生了巨大的变化。教育建筑也是一种建筑类型，同其他公共设施一样，呼应了不断变化的意识形态，将教室从"教师在前面教学转变为一排排的桌椅、圆桌、露天学校，以及技术推动下的校舍的瓦解"。[1]但是这些制度的意识形态不仅影响了那些传道授业者的教育方法，也影响了那些建筑师们，他们试图用砖块和灰泥，还有我们在本书中看到的，用混凝土来掩饰理论和实践。

墨西哥阿瓜斯卡连特斯州的Escuela Bancaria学校、Viettel学院和UDEP校园建筑都是非常流行的教育建筑的范例，强调了外部和内部空间之间的必要关系。在处理公共和私人空间时，强调了保护学习空间的重要性。因此，校园建筑像小城市一样汇聚在一起，有街道、公共广场、小路、坡道、楼梯等，促进了师生之间的自然接触和思想交流。

这三个项目所处的国家和环境不同，分别由三家建筑事务所设计；然而，它们有许多共性。众多元素交织在一起，吸引学生和导师来探索校园建筑：了解光影的相互作用；实体与空间；露天和带顶空间；本地化和全球化建筑设计方法；打造可持续发展项目，促进与当地自然和区域气候的和谐发展。

正如安藤忠雄所言，建筑对使用者负有道德责任，以确保建筑能够激励和保护我们[2]。既然这样，那教育建筑就必

"When you plan a school, do you say that you will have seven seminar rooms… or is it something that somehow has the quality of being a place in which you are inspired? To somehow talk there, and to receive a kind of feeling of talk? Could there be those spaces which have a fireplace? There could be a gallery instead of a corridor. The gallery is really the classroom of the students, where the boy who didn't quite get what the teacher said. Could talk to another boy, a boy who seems to have a different kind of ear, and they both could understand." - Louis I. Khan, *Conversations with Students* (1969)

The architectural typology for educational purposes has significantly changed in the last two centuries. It is a typology which, like other institutional buildings, responds to the changing ideologies in power, which has transformed classrooms from "frontal teaching to rows of tables and chairs, round tables, open-air schools, and the technology-driven dissolution of the schoolhouse"[1]. But these institutional ideologies not only affect the educational approach of those delivering it, but also those trying to conceal theories and practice with brick and mortar – or as we will see in this issue, with concrete.

The Escuela Bancaria Aguascalientes, Viettel Academy and UDEP campus are examples of very current approaches to educational architecture, highlighting the necessary relationship between outside and inside spaces; public and private spaces that reinforce the importance of protecting learning spaces. As a result, the campus buildings come together as small cities with their streets, public squares, paths, ramps, staircases, etc., promoting casual encounters and exchanging ideas between students and teachers.

The three projects are located in different countries and contexts and have been designed by three architectural practices; however, they share a number of commonalities. These elements, combined, invite students and tutors to explore the architecture of the campus, to learn about the interplay of lights and shadows; solids and voids; open-air and covered spaces; vernacular and global approaches to architecture; sustainable projects which create a harmonious atmosphere with the local nature and their regional climate.

As Tadao Ando argued, architecture has a moral responsibility with the user, to ensure that buildings inspire and protect us[2]. In this case, educational architecture must protect and inspire our education, creating spaces for more flexible pedagogies and learning encounters. By doing so, architecture will offer buildings which

须保护和激励教学，为更灵活的教学和学习机会创造空间。通过这种方式，建筑将提供楼宇，帮助我们"以完整的身体和精神上的存在去体验生活"[3]。在现实接触和社会互动已经近乎停止的时代[4]，分析这些教育综合体，得出了一个切中要害的结论：我们不能仅仅在虚拟平台上工作和学习，更需要感受建筑带来的学习体验。

墨西哥阿瓜斯卡连特斯州的银行商业学校，由Ignacio Urquiza、Bernardo Quinzaños＋Rodrigo Valenzuela Jerez、Camilo Moreno设计（122页），这是一个教育综合体，在这里你可以找到并体验到许多上述元素。正如这些专业人士在他们的项目描述中所述，他们项目的核心理念是"共存与自然交互，构建社会结构"。建筑师利用水平和垂直空间的交叉设计，构建了虚实空间，为学习行为、知识探索、师生聚会或自我反思提供了体验场所。

这个建筑综合体采用向内型设计，质朴的混凝土外观与花园、天井及垂直的混凝土结构形成对比，混凝土结构将室外的景色框了起来，与天空直接相连。其建筑风格大胆简约，将使用者的注意力集中在社交互动空间上。在某种程度上，受到前西班牙时期和殖民时期建筑类型的影响，开放空间与宏伟的结构并存，光影交错的建筑效果由此而生。这些建筑类型基于当地的气候、文化和建筑特点，促进了本土环境中的社会互动。

这个教育综合体规模较大，包括公共空间、楼梯和外露的方格天花板，其设计重视课堂之外发生的社会交流和教育作用。综合体便于室内外活动的开展，这些都是学校教学使命的核心。该校园建筑获得了墨西哥国内和国际上的认可，一举拿下了2019年第四届墨西哥城市建筑双年展的银奖[5]、芝加哥文艺协会和纽约建筑联盟颁发的奖项。

越南Viettel学院教育中心（108页）由Vo Trong Nghia建筑师事务所设计，距越南河内30多公里。受当地环境的影响，

help us to "experience ourselves as complete embodied and spiritual beings"[3]. At a time when physical encounters and social interactions have come to a halt[4], analysing these educational complexes introduces a very poignant conclusion: we cannot just work and learn on virtual platforms. We need to experience learning encounters facilitated by architecture.

The Banking and Commercial School, Aguascalientes, Mexico, designed by Ignacio Urquiza, Bernardo Quinzaños + Rodrigo Valenzuela Jerez, Camilo Moreno (p.122), is an educational complex where many of these elements can be found and experienced. As these professionals stated in their project description, at the core of their project was the idea of "coexistence and natural interaction, where social fabric is generated". This is achieved by a design that plays with the intersection of horizontal and vertical voids which generate solids and voids, inviting experiences of learning, of discovery, of getting together or reflecting on your own space.

The complex is inward looking, with an austere exterior of concrete which contrasts with the garden, the patios and the vertical concrete structures which frame the outside views, a straight connection with the sky. The style is bold and minimalistic, to focus the user's attention on the spaces created for social interaction. Somehow, the interplay of open spaces and monumental structures, as well as the light and shadow created as a result, can be traced back to prehispanic and colonial typologies which promote social interaction within a vernacular context, based on its climate, culture and architectural features.

The monumental scale of the educational complex, with its combination of public spaces, stairs and exposed waffle ceilings, give significance to the social exchange and educational role that takes place outside of the classroom: it frames it both as inside and outside activities, at the core of the educational mission of this school. The value of this campus has been recognised both nationally and internationally in 2019 with the Silver Medal IV Bienalof Architecture in Mexico City[5]; an award from The Chicago Athenaeum; and from the Architectural League of New York.

The Viettel Academy Educational Center (p.108), designed by Vo Trong Nghia Architects, is located 30km outside of Hanoi, Vietnam. This context drives this educational structure, creating a cooling micro climate

这座教育建筑要创造一种凉爽的微气候，对抗热带湿热的环境，让使用者在安静平和的室内外空间中进行学习。该综合体是一座设备齐全的独立建筑：12个建筑体块通过挑空空间、露台、水池和轻质屋顶相互连接，由人行道组成了如同错综复杂的迷宫一样的空间。花园、植被和公园打造了人与自然和谐交流的学习氛围。

这座大型建筑所选用的材料明确展现了其特质，它以当地砖块为建筑材料，其质朴的外观与混凝土道路形成了强烈反差，体现了乡村与当代建筑的趋势。屋顶采用了光和影的戏剧性效果，将公共区域打造得如同一个大型日晷，可以识别时间的变化。水池和小路似乎与不同的学习楼层并行，促进了人与环境的情感上的联系。砖面的不同纹理与每个立面上的光与影、颜色和不同的光泽、体量都发生了相互作用：学生在学习途中，就能感受到不断变化的环境。

内部庭院由附近的自然光线和小路围合而成，成为社交活动的场地，这里就像一个圆形剧场，在这里我们与他人结识，举办一些可以激发进一步的思考或社会互动的活动。这家年轻的建筑设计公司对光线、自然环境和当地材料拥有敏锐度，再结合现代化的设计，在越南国内和国际上得到长足发展也就不足为奇了。此项设计荣获了多个奖项，包括2019年的Dezeen设计奖。[6]

UDEP大学的新教学楼（136页）由Barclay&Crousse Architecture建筑事务所设计，位于秘鲁北部的皮乌拉。该项目是对该地区低收入和农村地区学生人数急剧增加所做出的回应。该项目营造了一种轻松的学习氛围，为学生的非正式的学习、偶然的接触和思想交流提供了场所。该设计以大型天窗和露天结构为中心，为室内空间提供阴凉和微风。室内空间由一个整体化的外观构成包围，作为一个加固的结构，保护11个建筑的综合体。

which defies the tropical and humid environment, and invites the user to embrace learning activities within its quiet and peaceful inside and outside spaces. The complex has been designed as a self-contained establishment: twelve buildings are interconnected by voids, terraces, water pools, and lightweight roofs, in a very intricate labyrinth composed of walkways. The gardens, vegetation and the park create a learning atmosphere in communion with nature.

This large-scale facility has achieved a clear identity thanks to its materiality, which reflects the vernacular and the contemporary trends in architecture by mixing local bricks, with a rustic appearance in contrast with the concrete that paves the pathways and highlights them. The roofs introduce a dramatic use of light and shadow, transforming public areas into large sundials, recognising the movement of time. The pools and the pathways seem to run a parallel with different levels of learning, promoting emotional connections with the place. The brick brings different textures, and plays with light and shade, colour, and different shines and volumes on each facade: ever-changing surroundings for students as they follow their learning pathway.

The inner courtyard, defined by the natural light and paths that surround it, becomes the scenario of social activity and movement, like an amphitheatre where we acknowledge the others, activities that spark further reflection or social interaction. It does not come as a surprise, then, that this young practice of architecture is thriving nationally and internationally thanks to their sensitive approach to light, natural and local material within a contemporary design. This has been recognised by several awards, including the 2019 Dezeen Award, for the Viettel Academy Educational Center[6].

The University Facilities UDEP (p.136), by Barclay & Crousse Architecture is located in Piura, North Peru. The project emerged as a response to a significant increase in the number of students from low-income and rural areas in the region. The project embraces a learning atmosphere which delivers on informal learning, casual encounters, and exchanging ideas. The design revolves around large skylights and open-air structures that bring shade and breeze, indoor spaces that are firmly framed by a monolithic exterior which act as a fortified structure, protecting the complex sum of eleven buildings.

This feeling of enclosed spaces is softened by the monumental cantilevered roofs that provide shade, and facilitate circulation; the voids between the roofs, creating very thin shafts, help to break up the monolithic structure of this mini city, providing light and natural ventilation. The overall concrete materiality of the

这种封闭空间的感觉被巨大的悬挑屋顶弱化了，屋顶提供了阴凉，也便于空气流通；屋顶之间的空隙构成了窄型通风口，有助于打破这个微型城市的整体化结构，提供了光线和自然通风。这个综合体的一体化混凝土质感在一些地方是断裂的，从而减小了建筑的密度。主立面可以过滤强光，抵御皮乌拉的干旱气候。植被非常稀疏，其他材料和颜色也少有使用，建筑关注混凝土的质地及它对光线的反应：与完全沐浴在阳光下的外观形成对比，室内空间像子宫一样，为使用者遮风挡雨。

Barclay&Crousse建筑事务所的建筑宣言伴随着对地方和人类健康之间的关系的关注而日渐丰富，他们始终关注时间、空间以及光线。这家设计实验室从发展中国家的具体情况出发，通过探索景观、气候和建筑之间的关系，挑战技术、日常应用和生活品质的概念，在全球背景下提供切实的信息。[7]这种致力于改善用户生活的、对空间设计所付出的努力，在2019年得到了认可，事务所获得Archdaily网站和Mies Crown Hall美洲大奖的一等奖。[8]

毫无疑问，这三个教育中心都体现出了设计空间为用户服务的道德义务，即在为用户提供空间的同时，将空间设计为学习背景，从中创造出各种复杂而有趣的可能性，满足多种空间体验，方便人与人的交往。这些建筑都展示了内外世界之间的灵活和谨慎的平衡，尊重各自的原生环境和传统。在某种程度上，尽管教育的意识形态不尽相同，但它们都具有某些特质，让我们想起宗教建筑，这是中世纪修道院的精髓，通过提供庇护所和开放空间来保护知识和智慧，让思想自由地驰骋。

complex is ruptured in places to remove its density, with the main facade acting like a filter for the very strong light and arid climate of Piura. The vegetation is very sparse, like the use of other materials and colors, focusing on the materiality of the concrete, and how its reaction to light: contrasting a completely sunbathed exterior, and a womb-like protected interior space, which gives shelter to the user.

Barclay & Crousse's architectural manifesto grows from a strong focus on the relationship between place and human wellbeing, paying special attention to time, space and light: a design laboratory that explores the bonds between landscape, climate and architecture, in order to challenge those notions of technology, usage, and quality of life that, from the specific conditions of developing countries, can inform and be pertinent in a global context[7]. This dedication to design spaces that improve the life of its users, has been recognised in 2019 with the First Prize from Archdaily and the Mies Crown Hall Americas Prize[8].

These three educational centers certainly demonstrate a moral obligation to design spaces that serve the user, that offer intricate and interesting possibilities to serve as a backdrop of learning, embracing a multiplicity of spaces and experiences to facilitate these encounters. They all demonstrate a flexible and careful balance between the inside and outside worlds, acknowledging their contexts and traditions. Somehow, and despite the different ideologies on education, they all share certain qualities that remind us of religious architecture, the essence of medieval monasteries which protected knowledge and wisdom by offering shelter and open spaces to enable the mind to run free.

1. Nick Axel, Bill Balaskas, Nikolaus Hirsch, Sofia Lemos, and Carolina Rito 'Editorial. Architectures of Education' (2019), https://www.e-flux.com/architecture/education/322661/editorial/
2. Tadao Ando, *Conversaciones Con Michael Auping* (Barcelona: Gustavo Gili, 2007 [2003]), p.15.
3. Juhani Pallasmaa, *The Eyes of the Skin: Architecture and the Senses* (Great Britain: John Wiley & Sons, 2007), p.11.
4. At the time of writing this article, the United Kingdom has implemented social distancing measures to tackle the Covid-19 pandemic (March 2020).
5. https://www.colegiodearquitectoscdmx.org/2019/10/09/ganadores-iv-bienal-de-arquitectura-de-la-ciudad-de-mexico/; https://archleague.org/centro-de-colaboracion-arquitectonica/
6. http://votrongnghia.com/company/
7. http://www.barclaycrousse.com/about
8. http://www.barclaycrousse.com/awards

Viettle学院教育中心

Viettel Academy Educational Center

Vo Trong Nghia Architects

由多层交通流线连接的12座建筑营造了阴凉处与室外空间

The twelve buildings are connected by multilevel circulations offering shades and outdoor spaces

Viettel学院教育中心位于距河内30km的Hoa Lac高科技园区，美丽的景观、湖泊和丰富的绿地空间环绕着建筑，形成凉爽的微气候。这家教育中心位于校园的中心位置，可以很方便地从居住区和其他教学楼进入。项目旨在为学员创造一个宁静平和的空间，让他们专注于自己的学习，远离熙攘喧嚣的城市生活。它将成为越南最大的移动网络运营商Viettel公司安排短期住宿和培训课程的地方。

教育中心由12个建筑体块构成，容纳了教室、会议室、大厅和办公室。主楼有四至五层楼高，而其余建筑体块仅有二至三层楼的高度。建筑体块被一个溢水池环绕着，不仅可以反射建筑和周围景观美丽的倒影，也有助于调节微气候。

体块由多层交通流线通道连接，如走廊、坡道和楼梯，形成了许多有趣的观景点和静谧的学习空间。由于河内的热带潮湿气候，建筑师设计了一个轻质混凝土屋顶遮盖住大部分半室外的空间，这个屋顶也作为空中走廊，有助于减少阳光直接辐射。

位于一层的花园在建筑体块之间交替布置，为学员营造了一种友好的氛围，让他们亲近自然。不同楼层的屋顶花园形成了一系列空中花园，为学生休息时的交流互动提供了一个放松的好去处。

建筑表面使用当地砖材，为整个项目打造了一种令人印象深刻的红砖立面。整体统一的立面散发出强烈的乡野气息。300~400mm厚的立面由中间留有空隙的两层砖墙构成，有利于隔绝日晒并减少能源消耗。砖墙立面成为体块周边活动的背景，为参加培训课程的学员创造了鲜活生动的回忆。红砖立面与周围绿色空间的结合，营造了建筑与自然之间的和谐连通。

Viettel Academy Educational Center is located in a Training Center Campus at Hoa Lac Hi-Tech Park, which is 30km away from Hanoi, Vietnam. Surrounded by beautiful landscape, lakes and abundant green space, the building has a cooling microclimate. The Educational Center is very accessible from the residential zone and other facilities due to its central location within the campus. The project aims to create a quiet and peaceful space for the trainees to focus on their studies, away from the hustle and bustle of city life. It will provide short-term accommodation and training courses of Viettel Corporation, Vietnam's largest mobile network operator.

The Educational Center consists of twelve blocks, accommodating classrooms, meeting rooms, halls, and offices. The main blocks are four to five stories whereas the rest are only two to three stories high. These blocks are surrounded by an overflow pool, which not only creates beautiful reflections of the build-

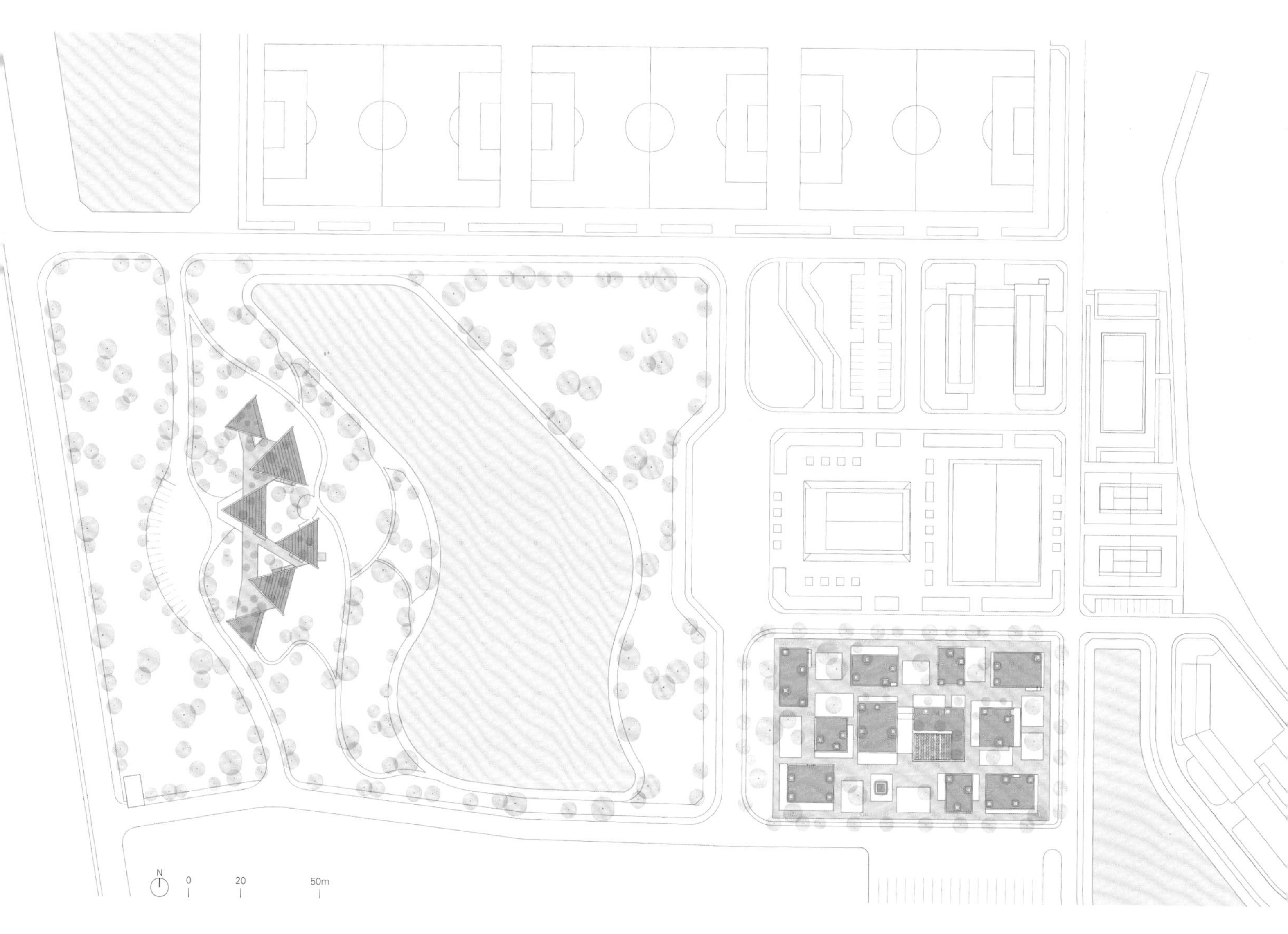
N
0
20
50m

ings and the surrounding landscape but also helps in regulating the microclimate.

The blocks are connected by multi-level circulation paths, such as corridors, ramps, and staircases. This offers many interesting views as well as various quiet areas for studying. Due to the humid tropical climate in Hanoi, a lightweight concrete roof is designed to cover the majority of semi-outdoor spaces, which also functions as sky walk. Besides this, the roof design helps to reduce direct radiation from sunlight.

The first-floor garden system is arranged alternately among the blocks, creating a friendly atmosphere for trainees and bringing them closer to nature. Roof gardens on different floors form a series of hanging gardens that provide students a relaxing space for interaction during breaks.

Local bricks are used for the building finishes, creating an impressive red-brick facade for the whole project. This monolithic facade exudes a strong and rustic presence. The 300~400mm-thick facade is made of two layers of brick wall with a void in-between for insulation to reduce energy use. The brick facades become the backdrop of activities that are taking place around the blocks, creating a vivid memory for the participants during the training course. The combination of the red brick facade with the surrounding green space creates a harmonious link between the building and nature.

项目名称：Viettel Academy Educational Center / 地点：Thach That, Ha Noi, Viet Nam / 建筑师：VTN Architects (Vo Trong Nghia Architects) / 总建筑师：Vo Trong Nghia
设计团队：Ngo Thuy Duong, Do Minh Thai, Do Huu Tam / 客户：Viettel Corporation / 承包商：Delta Corp / 功能：education facility
用地面积：9,026m² / 总建筑面积：2,651m² / 设计时间：2015 / 施工时间：2016.11 / 竣工时间：2018.3 / 摄影师：©Hiroyuki Oki (courtesy of the architect)

A–A' 剖面图 section A-A'

1. 教室 2. 保安室 3. 机械室 4. 接待大厅 5. 接待室 6. 办公室 7. 会议室 8. 电影院
9. 茶水间 10. 卫生间 11. 服务器室 12. 主管办公室 13. 副主管办公室 14. 图书室
1. classroom 2. security room 3. technical room 4. reception hall 5. reception room 6. office 7. meeting room
8. cinema 9. pantry room 10. WC 11. server room 12. director room 13. vice director room 14. library

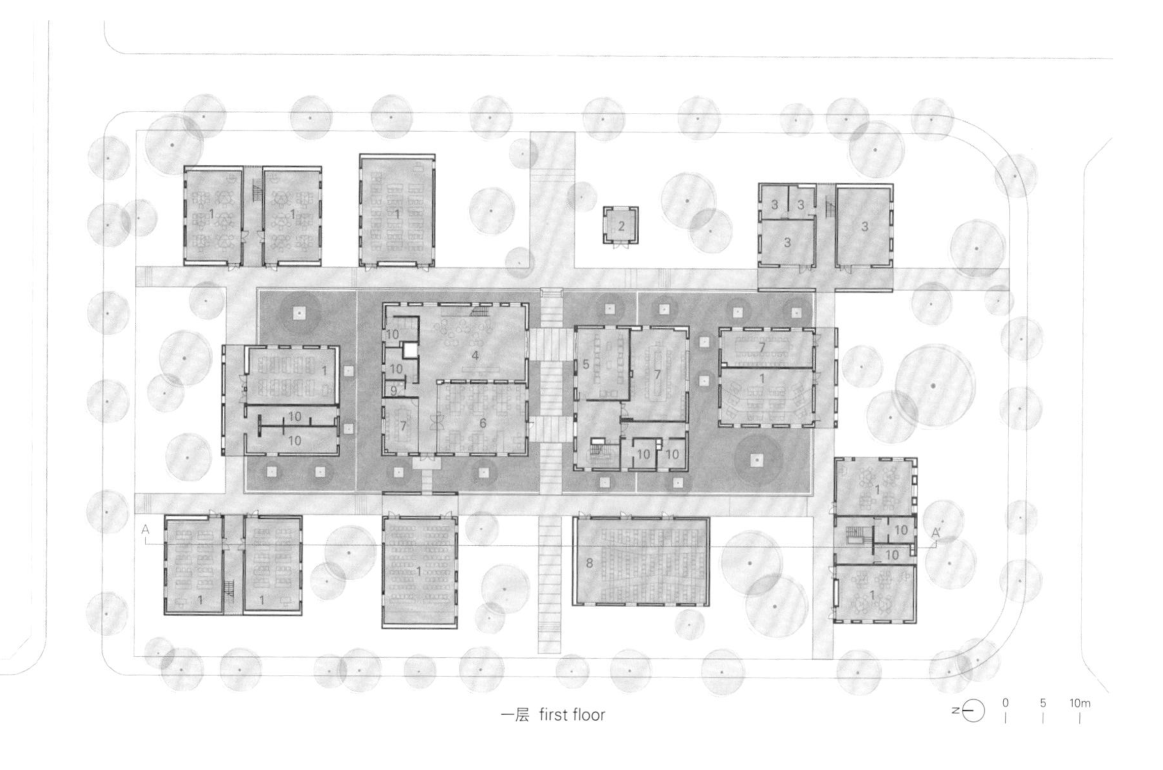

一层 first floor

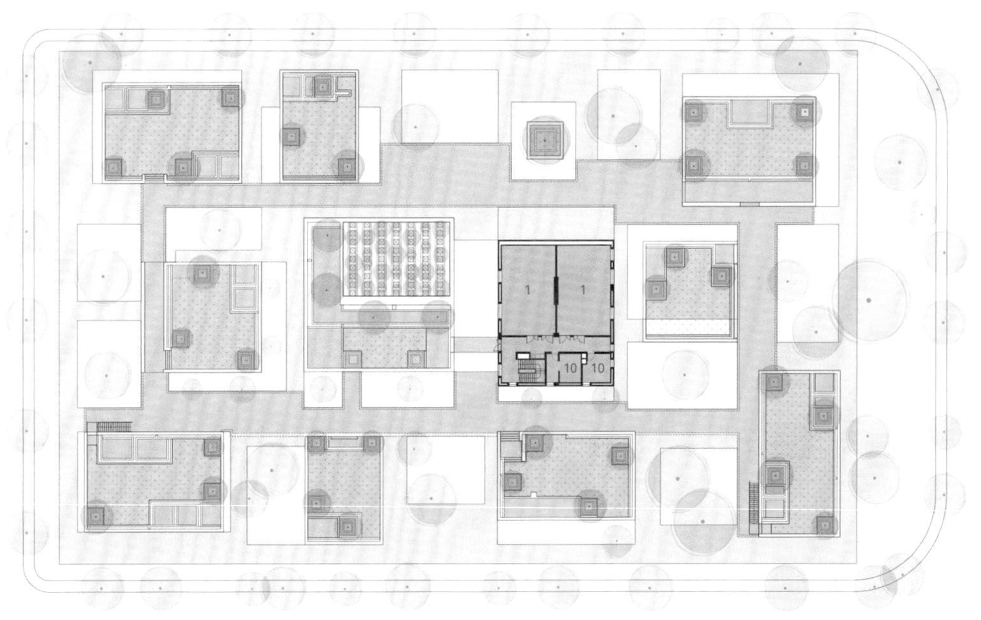

四层 fourth floor

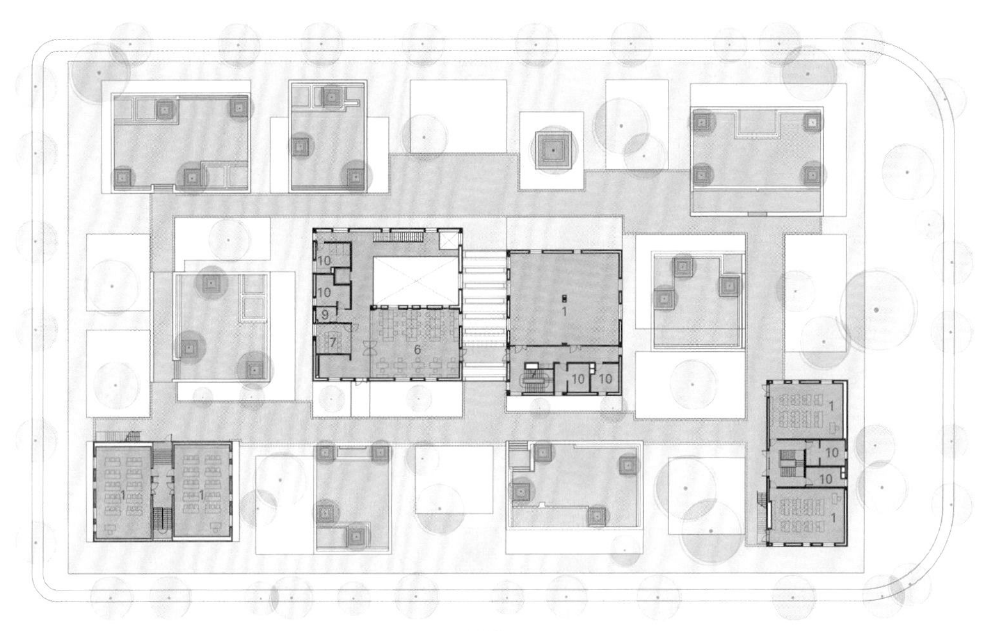

三层 third floor

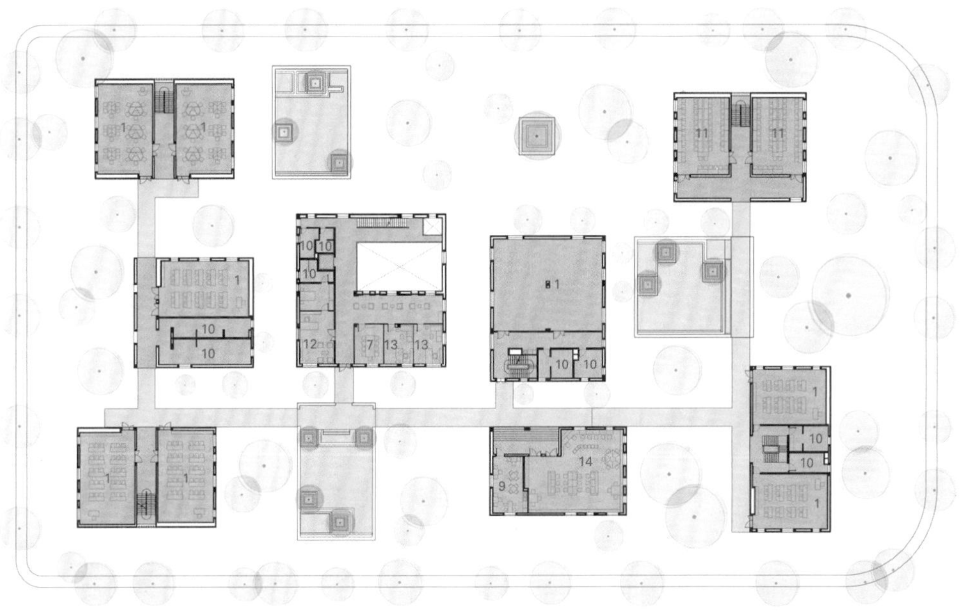

二层 second floor

CÔNG TY VỀ
ĐIỆN TỬ VÀ CƠ KHÍ
VIETTEL

阿瓜斯卡连特斯市银行商业学校

Banking and Commercial School, Aguascalientes

Ignacio Urquiza, Bernardo Quinzaños + Rodrigo Valenzuela Jerez, Camilo Moreno

实体与空间的交互作用促进了人与人之间的交流

The interplay of solid and void promoting exchange between people

本项目位于阿瓜斯卡连特斯市一个不断发展的城市扩张区域，也是一个以住宅区为主的郊区，该区域近年来已经得到了教育部门和商务部门的投资。

项目包括该国最古老的私立银行商业学校之——Escuela Bancaria y Comercial (EBC) 的高等教学大楼的开发建设。

这个教育机构的力量既在于它自身的价值，也在于它改造个体的能力，而个体又凭借其经验改造环境，丰富了整个社会。因此，教育建筑应当是一个促进教育和学习的平台，为增进使用者之间的智力交流提供空间。

该项目将学生体验置于设计的核心，呼应校园建筑的主题。其核心理念倡导将校园中最重要的空间设置为室外空间，提供使用者之间的共存和自然互动，便于举办集会和建立社交网。这个空间是一个聚会空间，其空间源于对构成校园的传统部分的排布方式，即具有逻辑性、实用性、功能性和灵活性。建筑由两层组成，横向和纵向的空间交叉，构建了唤起内心体验的空间感。实体和空间的相互作用形成两个大型的中央庭院、一个公共广场和一个供人沉思的花园。这些空间相互连接，被视为一个整体。廊道有助于增进交流、为学生指路，也消除了等级制度、允许思想的释放。

由于毗邻地块的未来发展无法预测，因此建筑师设计了一个内向型的校园。校园入口路线被设计成只有两个，分别来自主干道或停车场。教室位于校园的两个入口附近，这里是两条纵向交通流线的中心。每间教室都与内部花园连通，有助于提升专注力和学习效果；它们与外部走廊相连，使流通方式简单而有效。教室的外墙采用了预制模块的设计，目的在于优化组装，发挥技术操作优势。

即使在使用外部走廊的情况下，这种外墙也保证了教室内全神贯注的学习效果。多功能厅、礼堂均与通道相连，这使得教学社团能够开展活动，而不影响校园的正常教学运转。此外，这两个空间均与带顶的广场相连，可以举办多种具有不同特色的活动。

作为整个校园的核心，学习中心位于公共广场与内部花园之间。这使它既可以作为学生聚会的场所，又可作为专注工作的场地，实现了双重作用。

The project is located in a growing area of the city of Aguascalientes, in the predominantly residential outskirts of the urban sprawl, an area which in recent years has seen investment from the educational and commercial sector.

The project consisted of the development of a building for higher education, of one of the oldest private institutions of the country, the Banking and Commercial School – Escuela Bancaria y Comercial (EBC).

The strength of the institution lies in its values, as well as in its capacity to transform the individual, who goes on to transform the environment and enriches society as a whole with its experience. Therefore educational architecture should act as a platform that promotes the act of educating and learning by proposing spaces that encourage intellectual exchanges among its users.

The project puts the student experience at the heart of the design, to address the theme of the campus. The central concept supports the idea that the most important space on a campus is one that affords the coexistence and natural interaction of users – that space outside the classroom, where meetings take place and the social fabric is generated. This space is a meeting space, understood as the void resulting from the most logical, practical, functional and flexible arrangement of the traditional parts that make up the campus.

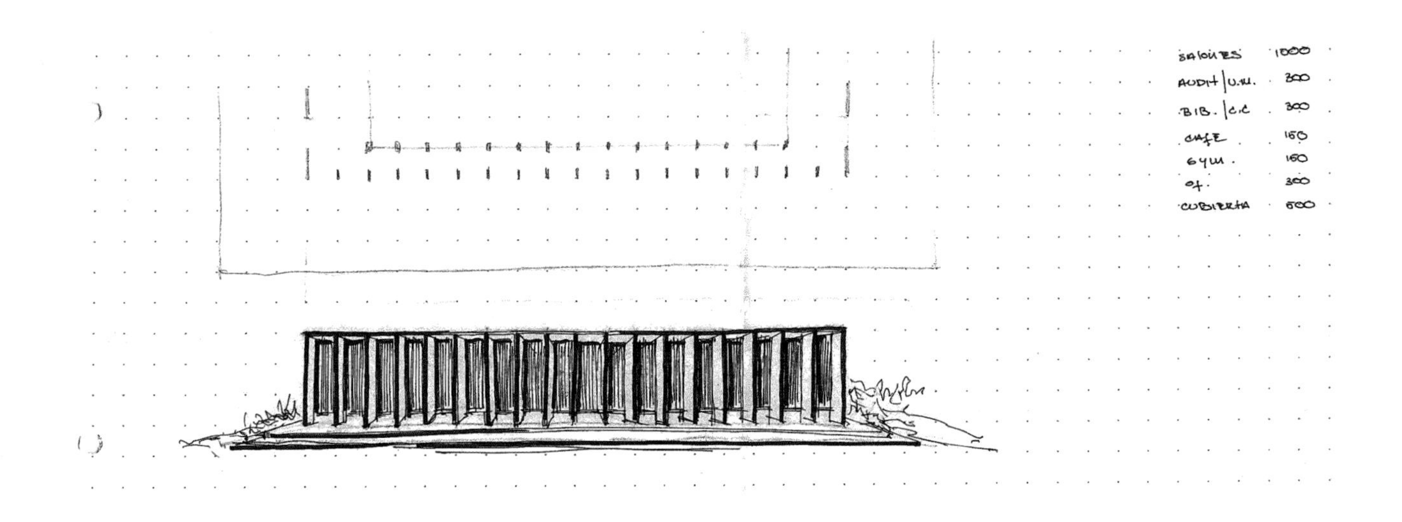

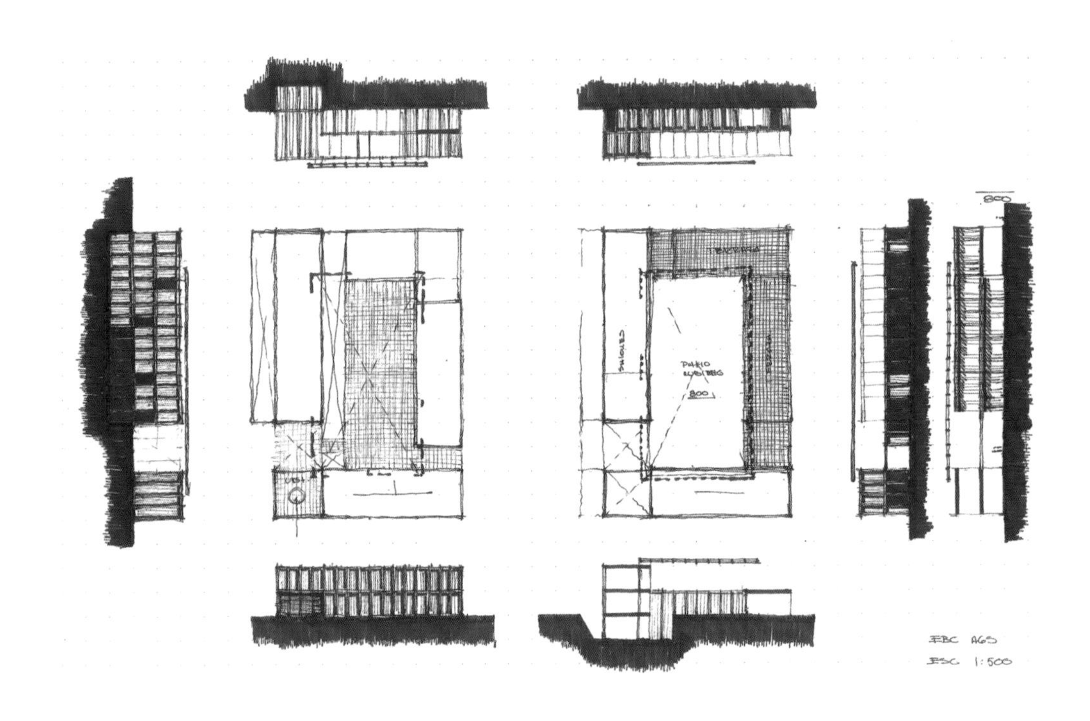

©Ignacio Urquiza Seoane (courtesy of the architect)

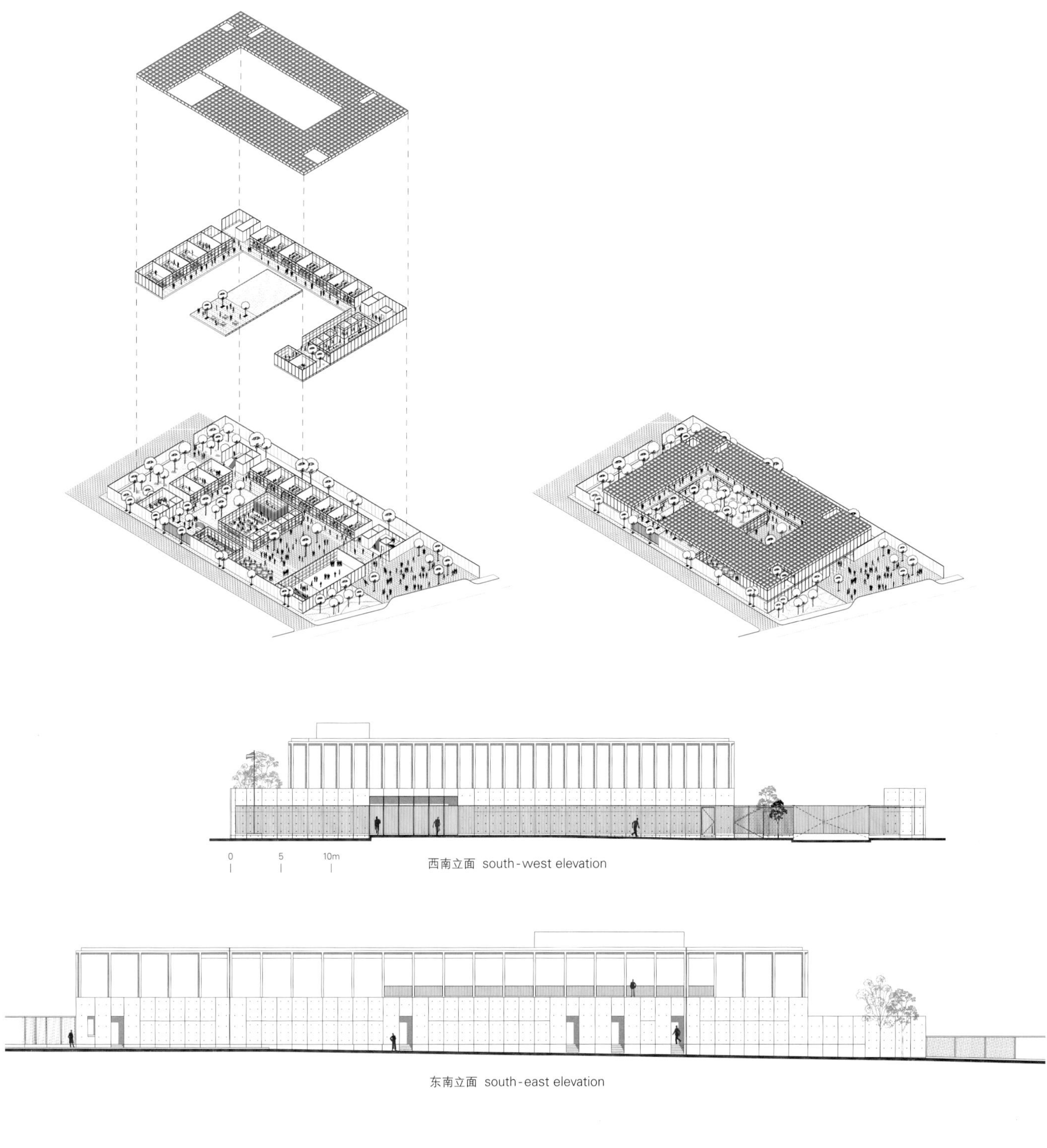

西南立面 south-west elevation

东南立面 south-east elevation

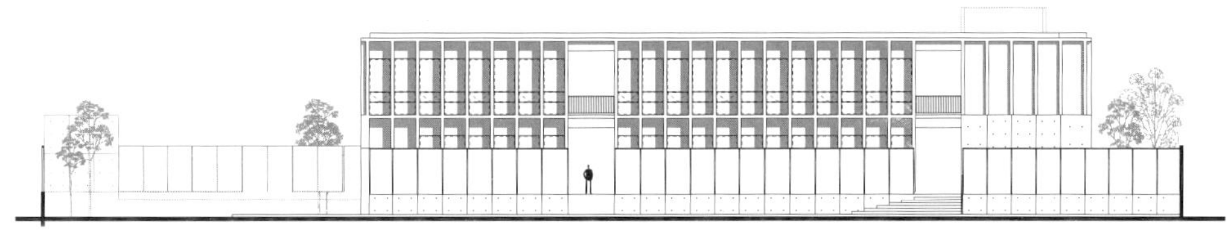

东北立面 north-east elevation

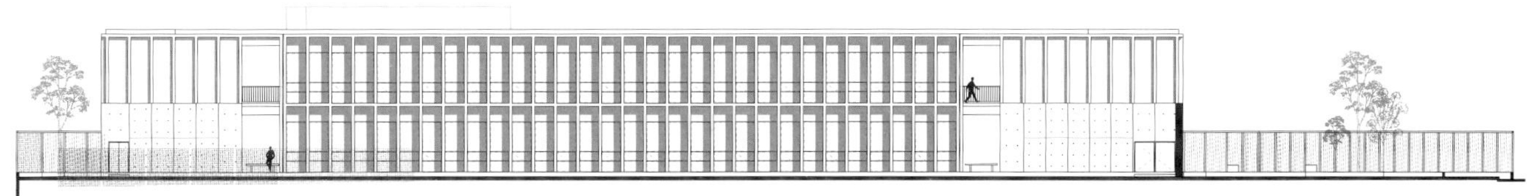

西北立面 north-west elevation

The building is composed of two levels, crossed by horizontal and vertical voids that generate spatiality arousing inner experience. The interplay of solids and voids makes two large central courtyards, a public square and a contemplative garden. These spaces are connected and are understood as a whole. The street promotes exchange, guides the student, eliminates hierarchies and allows thought to be released.

The unknown future development of the adjoining lots guided the creation of an inward-looking campus. The access routes have been designed to have only two control points, from the main avenue or from the parking lots. The classrooms are located near the two entrances of the campus, where two cores of vertical circulation have been located. Each room is related to an interior garden, which enhances concentration and study; these are connected to an exterior corridor that organizes the circulation in a simple and efficient way. As an enclosure of the classrooms, a skin based on prefabricated modules has been designed, with the aim of optimizing its assembly and obtaining technical operating advantages.

This skin guarantees good concentration inside the rooms even when the exterior corridors are used. The multipurpose room and the auditorium are linked to the access route; this

allows the educational community to carry out activities in a way which does not interrupt the normal functioning of the campus. In addition, these two spaces are linked to the covered plaza, which allows multiple events of different characteristics.

The learning center is the heart of the campus and is located between the public plaza and the interior garden. This reflects its double condition as a meeting place between the students and at the same time as a place of work and concentration.

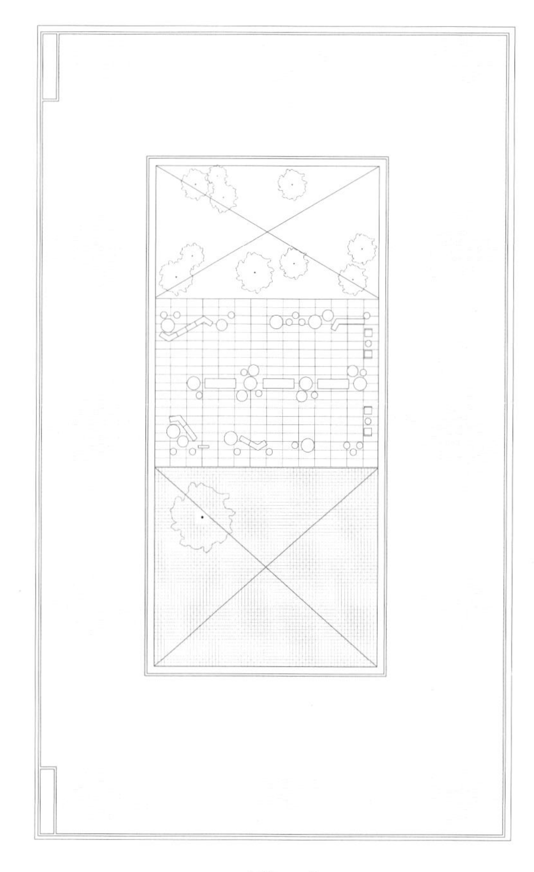

屋顶 roof

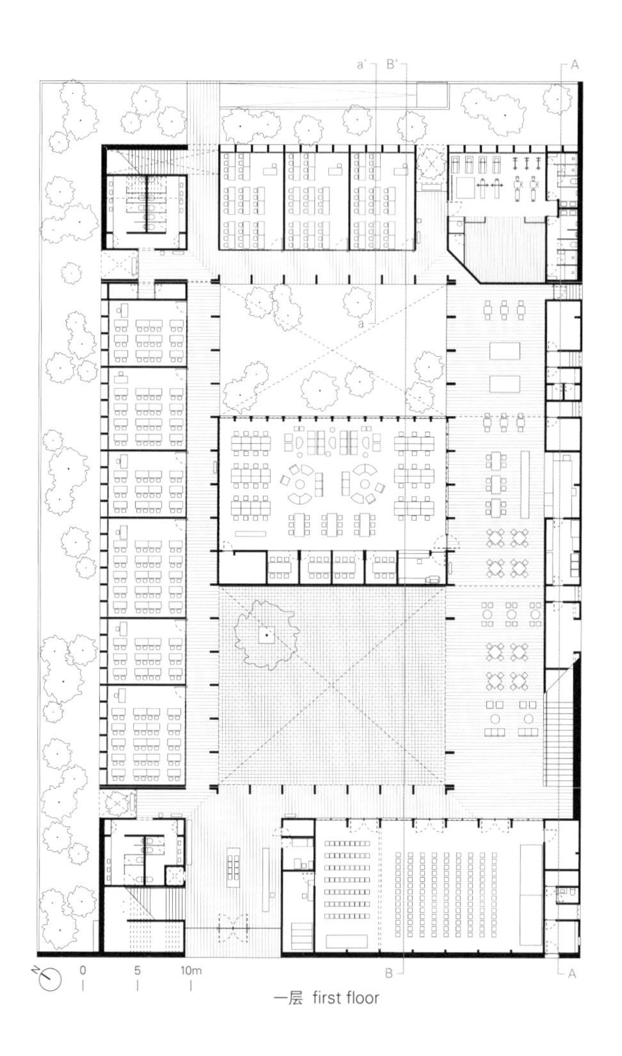

一层 first floor

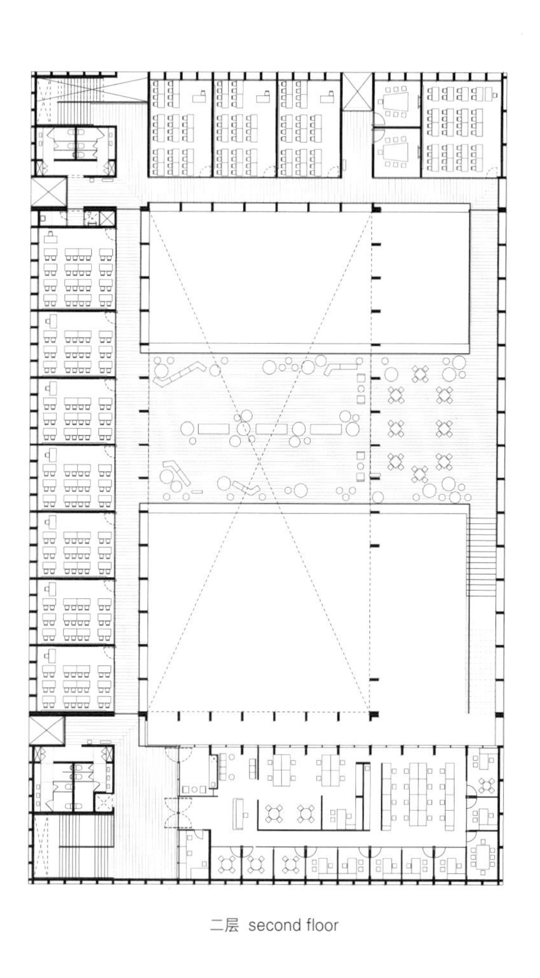

二层 second floor

项目名称：Escuela Bancaria y Comercial (EBC), Aguascalientes / 地点：Aguascalientes, México
建筑师：Ignacio Urquiza, Bernardo Quinzaños, Centro de Colaboración Arquitectónica in collaboration with Rodrigo Valenzuela Jerez and Camilo Moreno
设计团队：León Chávez, Fabiola Antonini, Daniel Moreno, Anais Casas, Héctor René Campagna / 结构：Ricardo Camacho / 工程：AKF Group
灯光：Inlight / 景观：Genfor Landscaping-Tanya Eguiluz / 室内设计：Taller Leticia Serrano / 开发与施工：Terral-Begoña Man zano, Jerónimo Prieto
客户：EBC / 面积：4,475m² / 竣工时间：2018 / 摄影师：©Onnis Luque (courtesy of the architect), except as noted

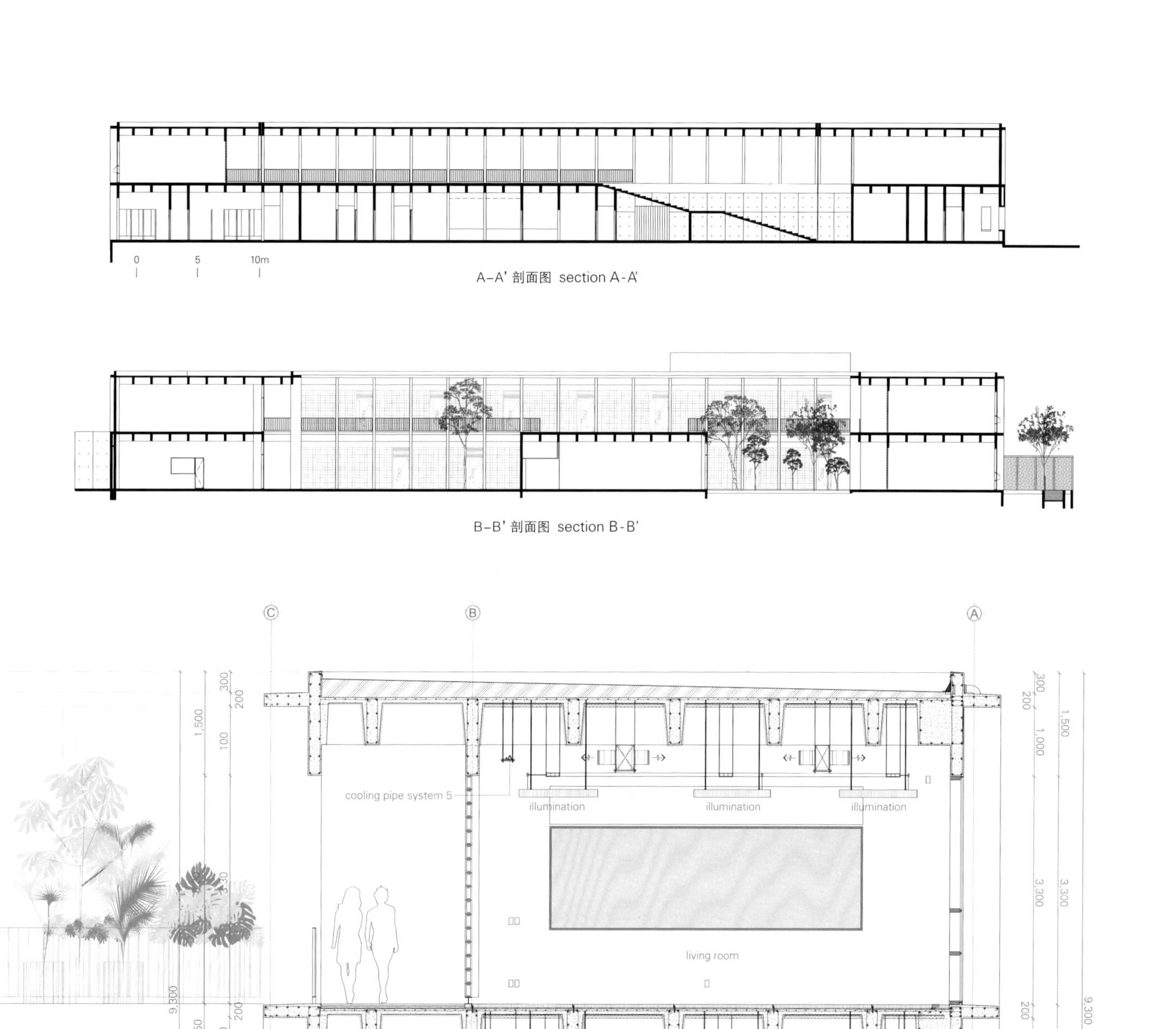

A–A' 剖面图 section A-A'

B–B' 剖面图 section B-B'

a–a' 剖面详图 detail a-a'

UDEP的大学设施
University Facilities UDEP

Barclay & Crousse Architecture

秘鲁沙漠地区11个建筑体量之间的适应气候的公共空间
Climate responsive communal spaces between eleven volumes in Peruvian desert regions

皮乌拉大学（UDEP）校园位于利马北部近1000km，坐落于皮乌拉市的城市网格之中的一大片土地上。周边地区是赤道附近典型的干燥森林，主要由生长在沙土中的角豆树组成。最近，该大学获得了一笔公共赠款，以便招收更多来自低收入地区的农村学生，因此迫切需要一座新馆，以容纳不断增加的学生人数。

对于建筑师来说，该项目的主要目标是营造一种学习氛围，而不是打造一个建筑类型或型态。大楼应具有促进非正式学习的功能：鼓励学生之间、学生和教师之间的轻松交往，促进思想交流，这一点尤为重要。

为了实现这一目标，在秘鲁北部这样一个日照时间长、炎热和干燥的沙漠气候中创造一个舒适的地带显得至关重要。就像干燥的森林适应干旱地区的生活一样，在70m×70m见方的建筑范围内，这些露天空间以同样的方式，即创造阴凉和带来凉爽的微风，鼓励学生们进行学术交流。

从外部看，建筑似乎是一个整体。但是一进入空间，参观者就会发现11座独立的建筑，每一座都有两到三层楼高，在宽敞的悬挑屋顶下，每一个悬挑屋顶都为多个聚会空间和交通流线空间提供了阴影。这些屋顶之间留有空隙，确保足够的自然通风和照明。这些缝隙创造了日晷的完美效果，一天中，太阳在地板和墙壁上投射出各种光线和生动的阴影。

11座建筑被设计在一个合理的方形布局中，同时在它们之间创造出了间隙和如同迷宫一般的空间。这些间隙产生了一系列不常被留意的空间，可用于聚会、休息和散步。建筑的多个入口使得学生可以通过步行，从校园的一个地方到另一个地方，鼓励了人际交流和接触。

该建筑的立面则根据所处热带环境的朝向配备了垂直百叶和预制格架。

The Universidad de Piura (UDEP) campus is located on a huge site, which nowadays lies within the urban grid of the city of Piura, nearly about 1,000km north of Lima. The surrounding area is a very interesting example of equatorial dry forest, consisting mainly of carob trees growing in sandy soil. Recently, the university was offered a public grant in order to admit more students from low-income, rural areas, and a new pavilion was urgently needed to accommodate this increase in the student population.

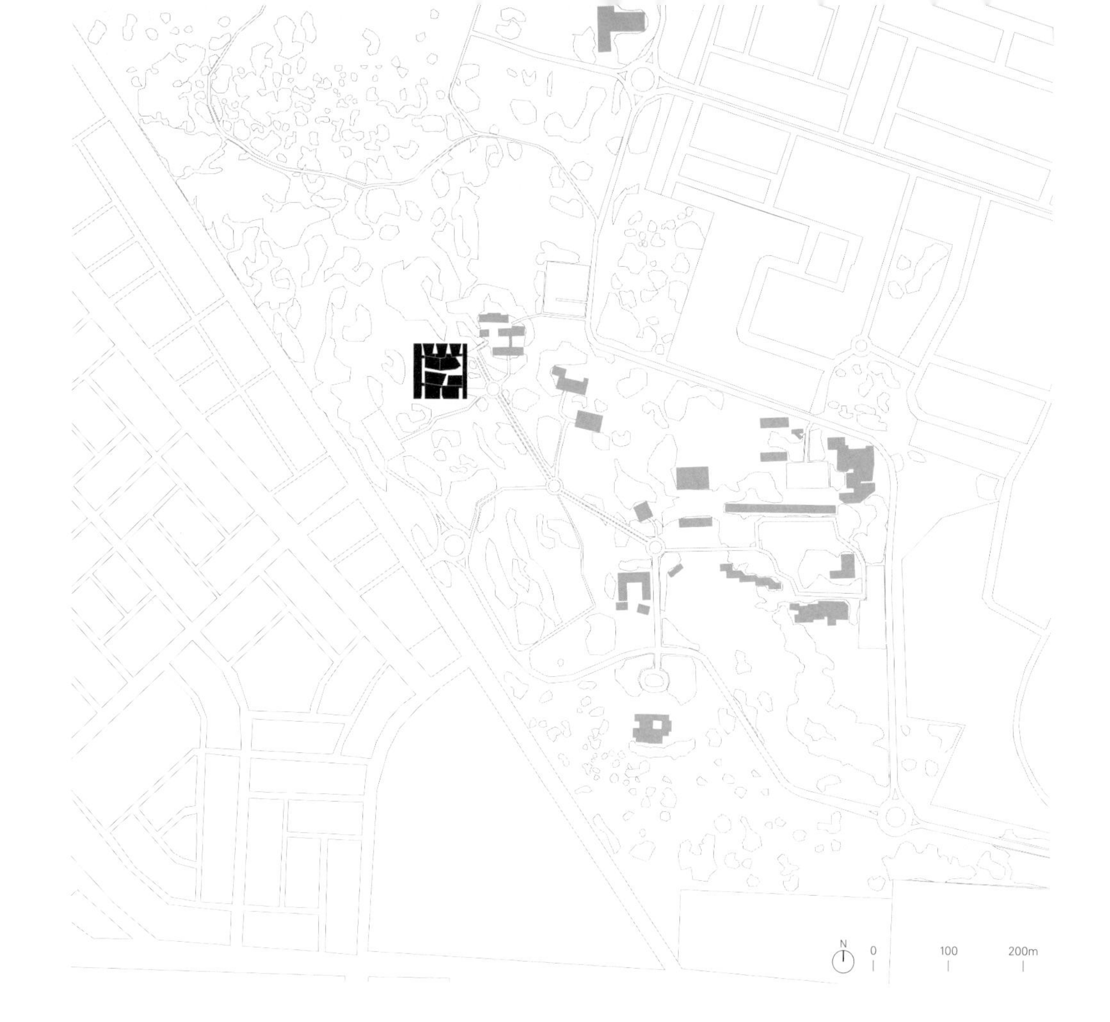
N
0
100
200m

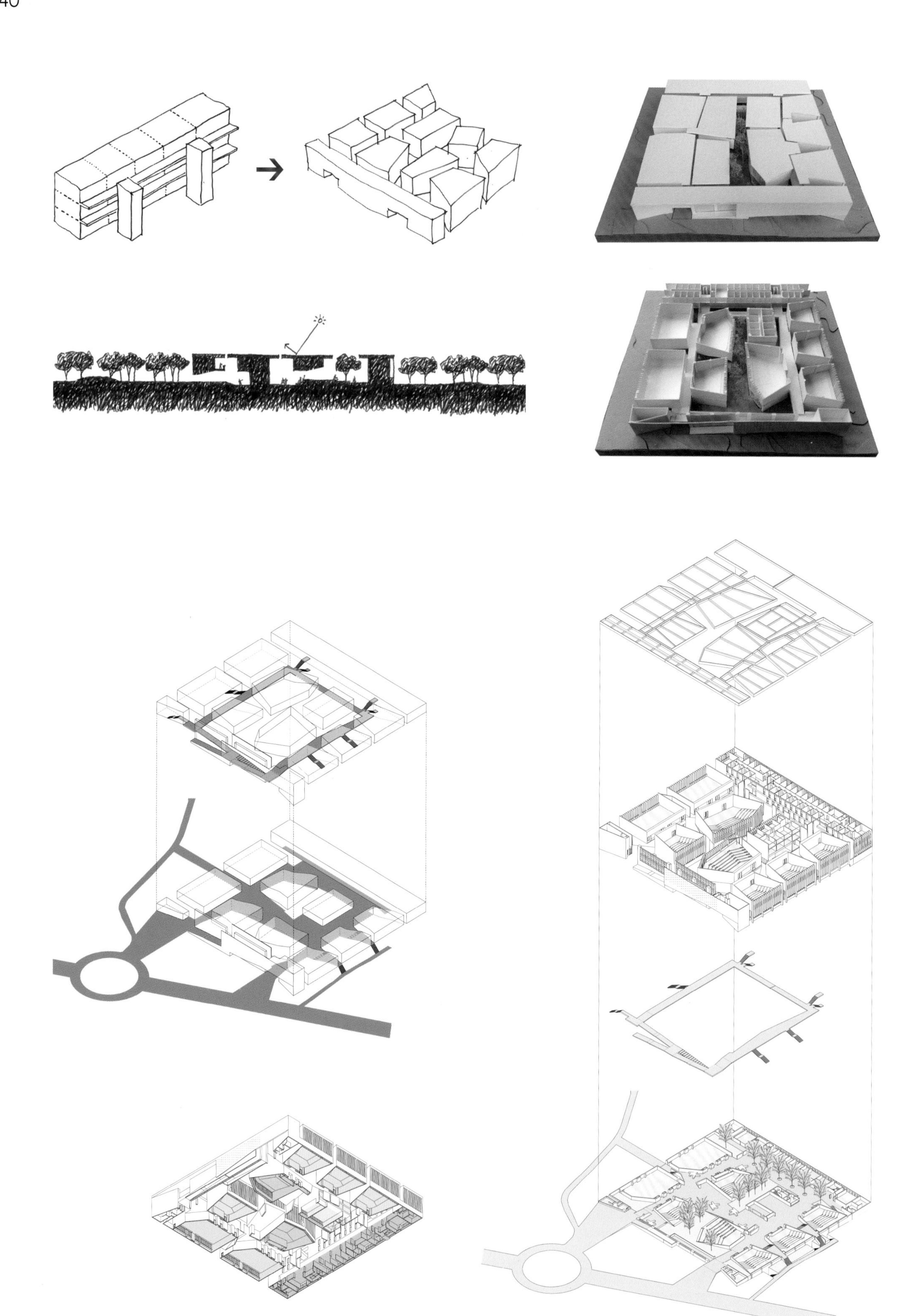

1. 入口
2. 阶梯教室
3. 工作台办公区
4. 卫生间
5. 工作空间
6. 主要区域
7. 咨询服务室
8. 办公室
9. 会议室
10. 工作室
11. 接待处
12. 行政办公室

1. entrance
2. lecture hall
3. table area
4. toilet
5. working room
6. master area
7. counseling
8. offices
9. conference room
10. atelier
11. reception
12. administration office

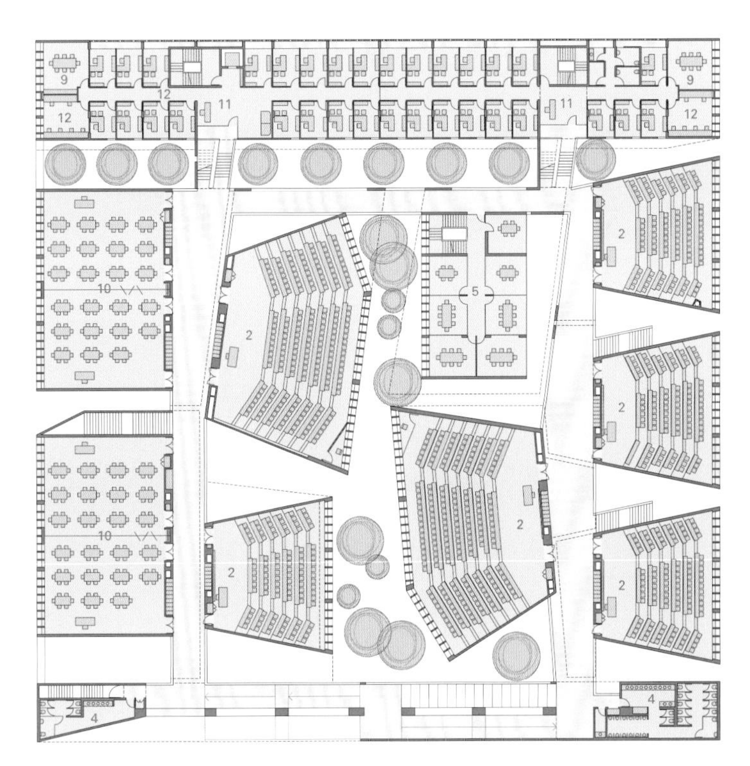

二层 second floor

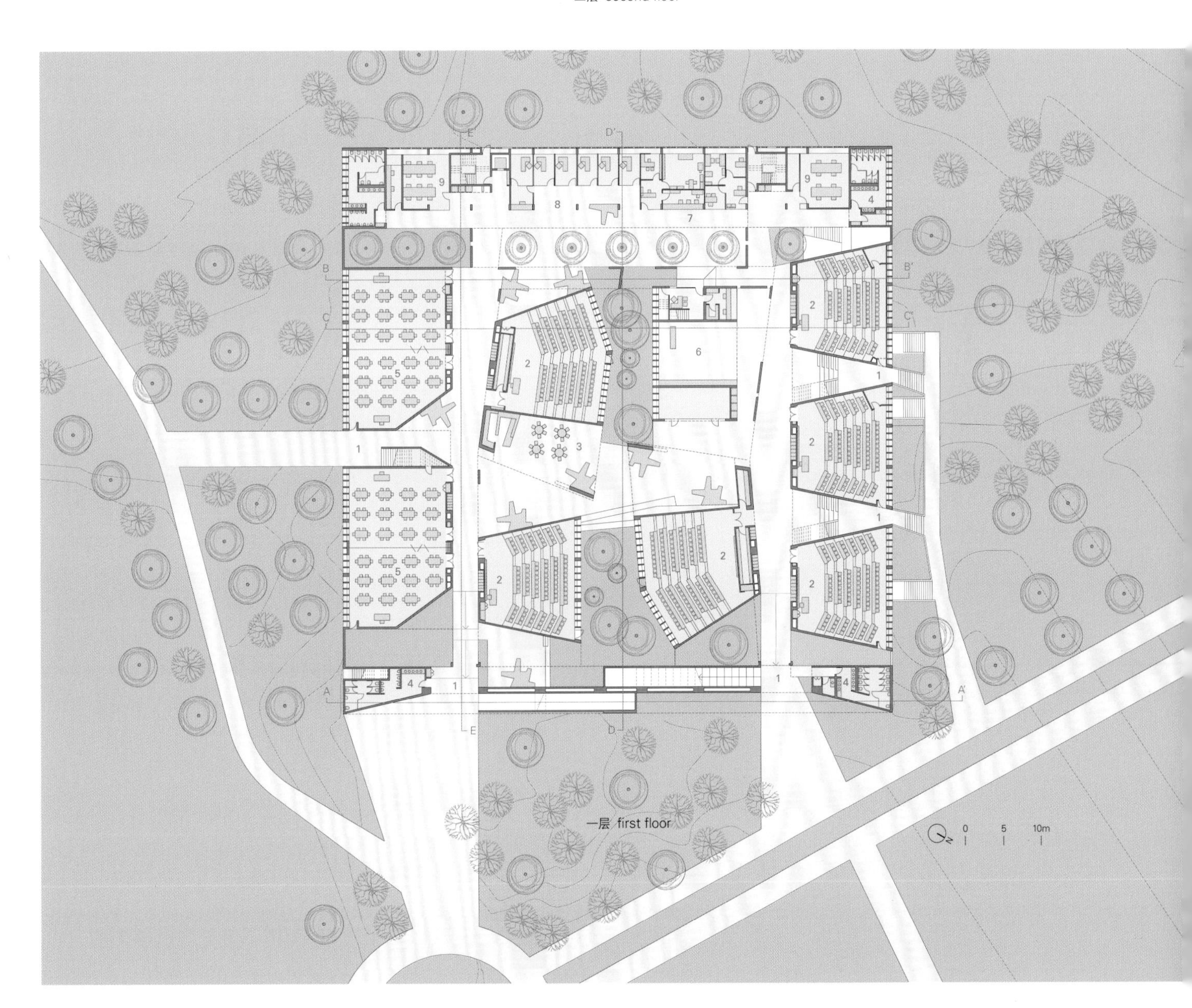

一层 first floor

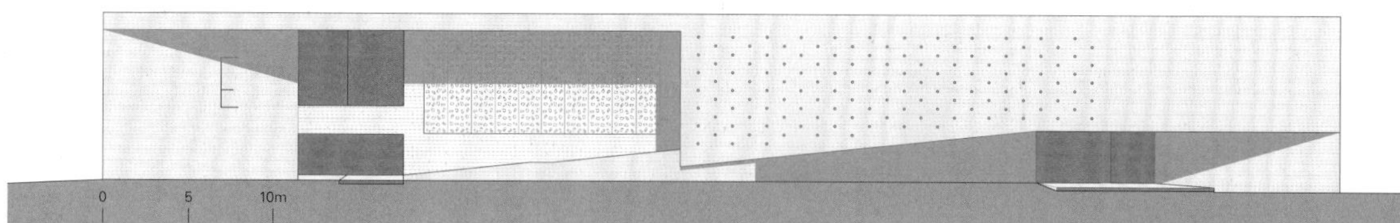

东立面 east elevation

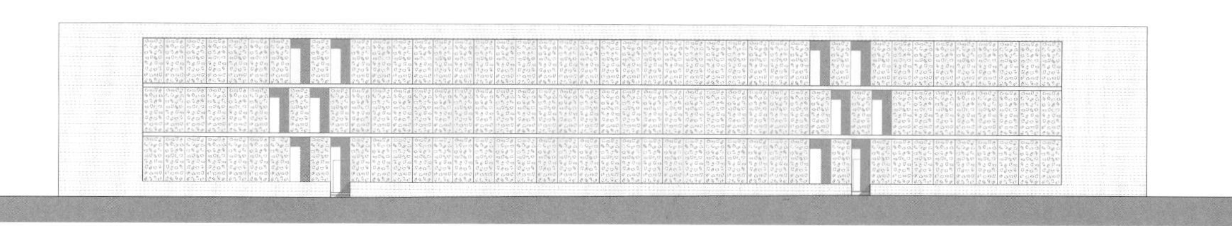

西立面 west elevation

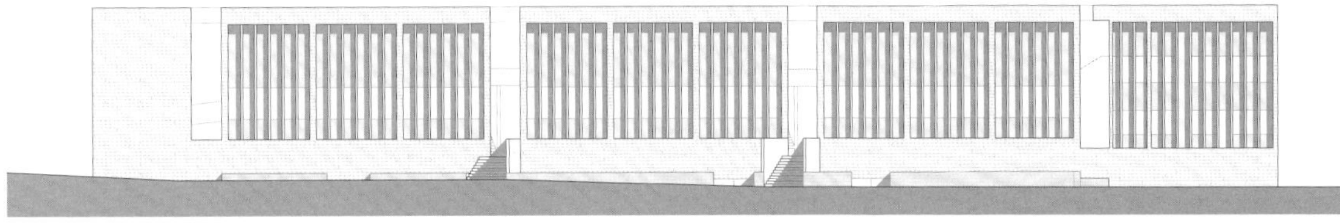

北立面 north elevation

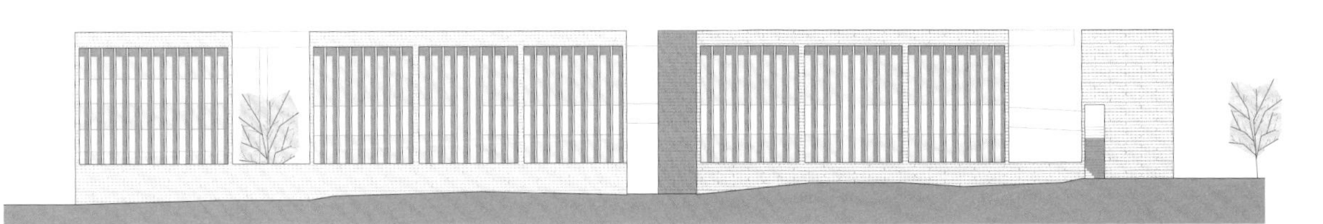

南立面 south elevation

For the architects, the main goal of the project was to create a learning atmosphere more than an architectonic type or shape. It was considered important that the building should be capable of fostering informal learning: a friendly environment which encourages casual encounters between students and between students and teachers, contributing to the exchange of ideas.

To achieve this, creating a comfortable zone in the permanently sunny, hot and dry climate of the Peruvian northern desert was of key importance. The open-air spaces, within the geometric 70x70m limits of the building, nurture the academic life of the students in the same way that the dry forest supports life in such an arid place: by creating shade and allowing for cooling breezes.

From the exterior the building appears monolithic but, once inside the space, the visitor discovers a group of 11 independent buildings, each two or three levels high, under generous cantilevered roofs that emerge from each one to provide shadow over multiple gathering and circulation spaces. These roofs leave gaps between them, ensuring adequate natural ventilation and lighting underneath. The gaps create the effect of a sundial, as over the course of the day, the sun casts light and dramatic shadows variously over floors and walls.

The eleven buildings are placed in a rational, square-shaped arrangement, which at the same time creates spaces between them that are interstitial and labyrinthine. These intermediary spaces generate a series of unattended possibilities for gathering, resting and strolling. Multiple accesses to the building complex encourage students to walk across the campus when moving from one place to another, encouraging circulation and encounters.

The facades are equipped with vertical louvers and prefab trellis depending on the orientation in the tropical setting.

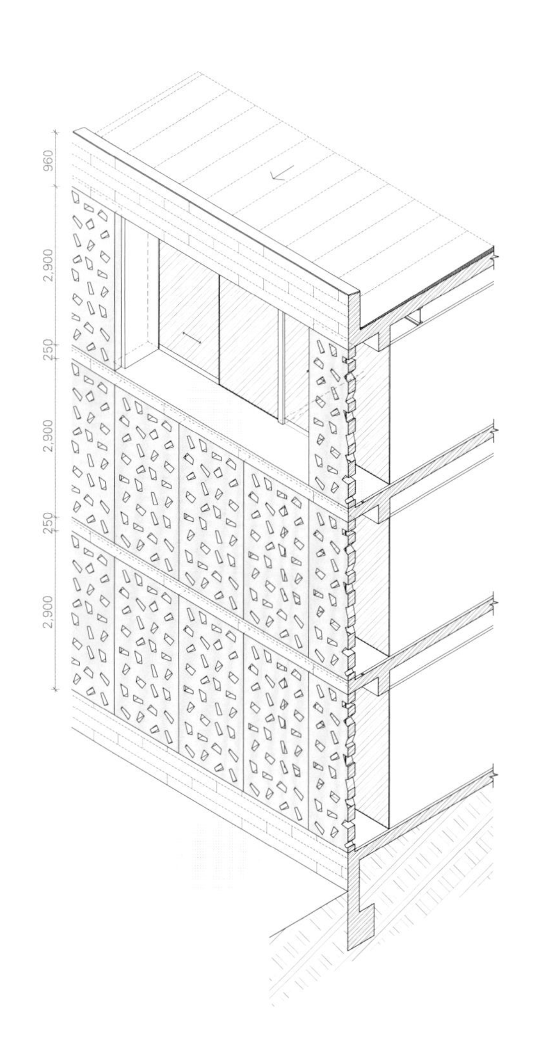

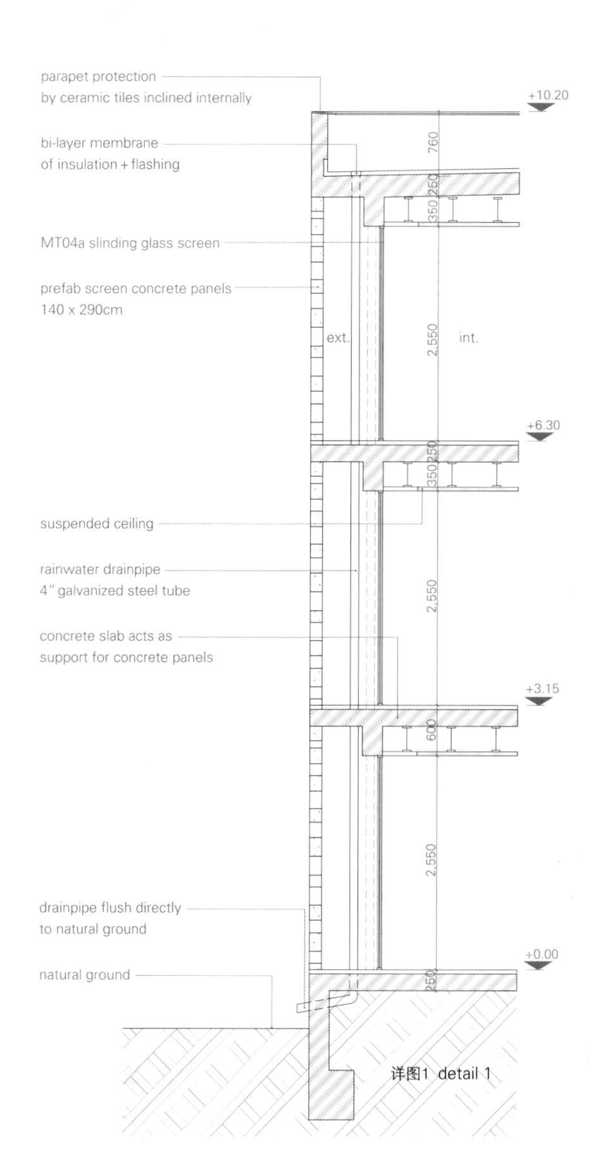

详图1 detail 1

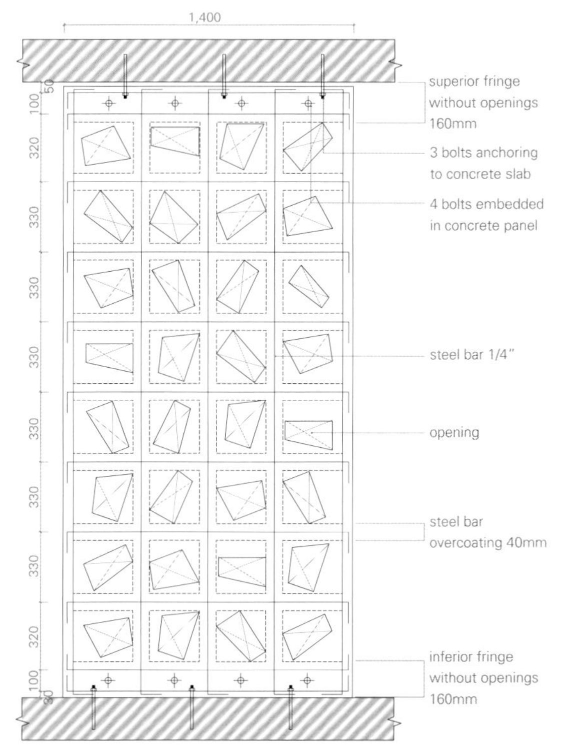

墙体部分立面 wall partial elevation

模板模块 formwork module

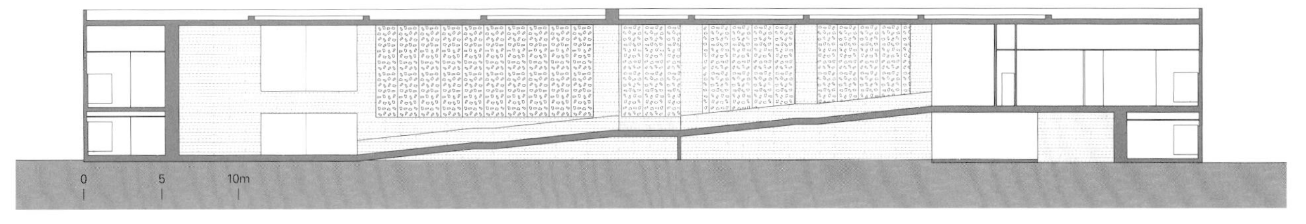

A–A' 剖面图 section A-A'

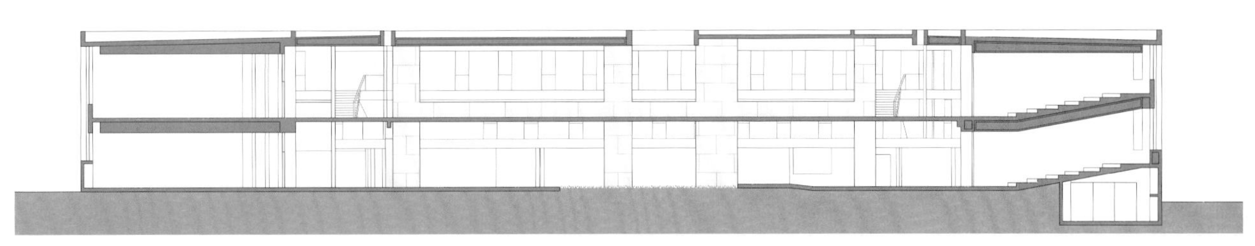

B–B' 剖面图 section B-B'

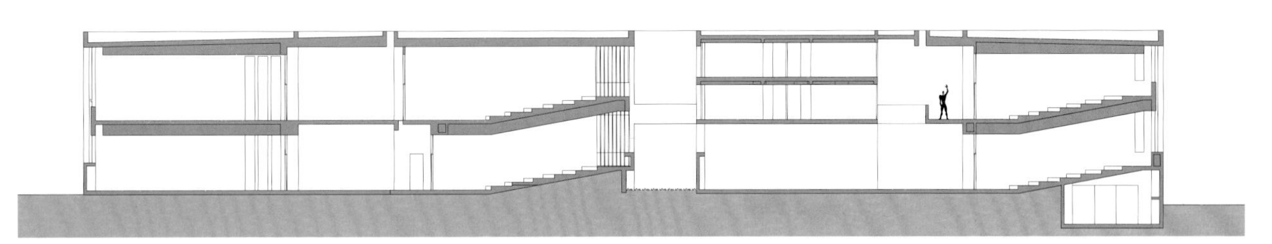

C–C' 剖面图 section C-C'

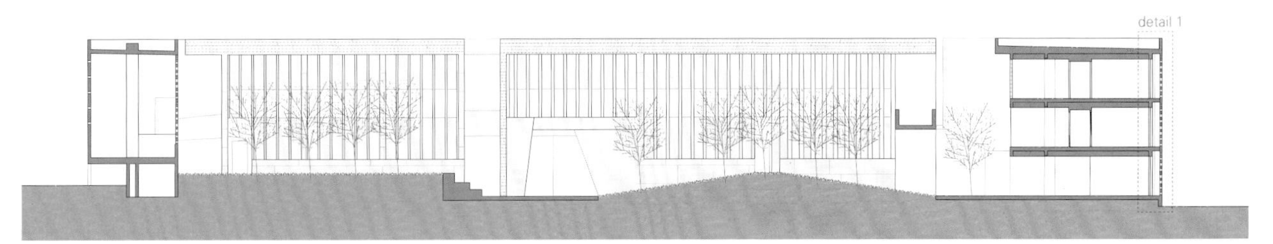

D–D' 剖面图 section D-D'

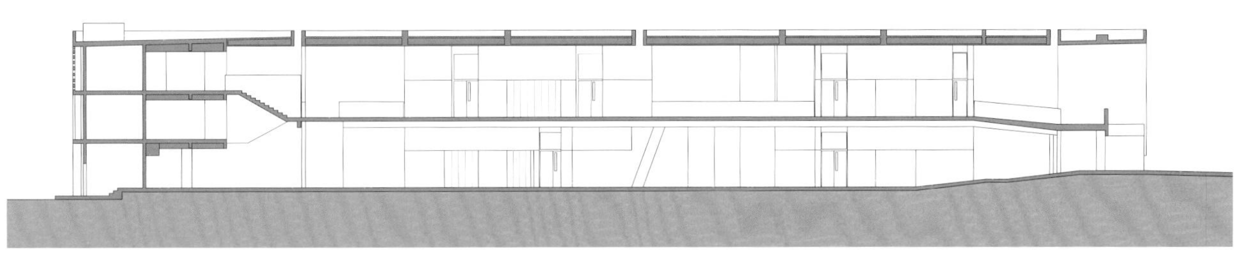

E–E' 剖面图 section E-E'

项目名称：University Facilities UDEP / 地点：University of Piura Campus, Piura, Perú / 建筑师：Barclay & Crousse Architecture – Sandra Barclay, Jean Pierre Crousse
助理：David Leininger / 工程：structural design – Higashi Ingenieros; security, life safety and fire protection – ESSAC Engineering; electrical engineering – MQ & Ingenieros Asociados; plumbing – Equipo "G" S.A. / 技术管理：SICG SAC / UDEP基础设施部门：Ing. Gonzalo Escajadillo / 承包商：Huarcaya Construcción – Ingeniería / 客户：Universidad de Piura (UDEP) / 建筑面积：9,400m² / 材料：exposed concrete, polished concrete, painted concrete (inner walls), on-site prefab concrete shading panels, exterior floors in stamped cement and cobble stones, interior floors in vinyl, joinery in painted wood, glazed surfaces in frameless tempered glass / 设计时间：2015.2—2015.5 / 施工时间：2015.6—2016.3
摄影师：©Cristobal Palma

SALIDA

中国建筑的新传统

A New Tr
in Chinese A

博物馆等重要的公共建筑和代表性建筑的设计，向建筑师提出了挑战，促使他们去寻找一种策略来应对创新与传统之间尚未解决的辩证关系。在中国，大多数现代建筑都在寻找一种当代国际建筑行业与当地文化遗产之间的可以接受的平衡。

有时，在新建筑中，我们可能会发现与中国传统特色完全相同的复制品，它们想要延续过去的设计语言，仅仅用混凝土和玻璃代替了旧的木头和砖块。而其他时候，我们发现巨大的建筑展示了解决方案和创新材料的丰富性，但却被归类为仅仅是使用了现代技术和设计语言，即一个属于全球建筑环境的现代化展示，并未考虑与当地历史的关系。

The design of such important public and representative buildings as museums challenges the architect to find a strategy to face the unresolved dialectic between innovation and tradition. In China, the majority of recent architecture struggles to find an acceptable balance between the contemporary and international construction industry, and the local cultural heritage.

At times, in new buildings we may find exact replicas of the Chinese traditional features: they intend to keep a continuity of language with the past, just substituting the old wood and bricks with concrete and glass. Other times, we find enormous buildings displaying richness of solutions and innovative materials, but grouped in only by the use of modern techniques and design language: a show of modernity that belongs to the global architecture environment, leaving apart any relation with local history.

adition
rchitecture

这些博物馆的共同主题是试图解决传统与创新之间的辩证关系，既避免表现出理想化的“中式风格”的形式特征，又避免仅仅成为当代全球建筑设计事务所的新案例。

乍一看，选定的项目似乎只是现代的混凝土块，应用某种创新的设计语言，植入普通的城市景观，并与之构建对话。通过梳理每个项目的外部空间，尤其是内部空间，我们可以更清晰地了解到哪些特征主导着设计的选择，以及空间品质如何与更广泛的建筑主题建立联系，其中许多主题都归属于中国的文化身份。

The main theme shared by these museums is the attempt to solve the dialectic between tradition and innovation, in such a way as to avoid the representation of formal features belonging to an idealized "Chinese style"; and to avoid being only a new example of the contemporary global construction practice.
At first sight, the selected projects may seem just modern concrete blocks, with a certain innovative language, inserted in a general urban landscape and setting a dialogue with it. By going through each project's exterior and above all interior space, it becomes clearer which features are leading the design choices, and how their spatial qualities do establish a link with a broader scope of architectural subjects, many of which belong to the Chinese cultural identity.

中国建筑的新传统
A New Tradition in Chinese Architecture

Andrea Giannotti

在人类知识和活动的许多领域，譬如多样化的文学和经济，再譬如遥远的农业和政治，围绕创新和传统概念的辩论仍然存在，在某些领域甚至非常活跃。它们的纯粹定义是不确定的，因为它碰巧是在不同的文化和空间、不同的时间所遇到的抽象概念。

如果我们试图从无数的现实情况中抽象化，给它们一个粗略的意义，我们就会说，传统是来自过去的方式和方法，甚至是某段没有记载的历史，我们现在正用它来处理一系列当今问题；另一方面，创新是我们对正面临的一系列问题所采用的前所未有、全新的方式或方法。

这两个概念相互交织，甚至可能重叠。许多评论家注意到，没有创新，传统就不可能存在，换句话说，这个观察结果仍然是有效的。同样，一个创新的解决方案也有可能成为传统的一部分，一个传统的特征可能会得到创新，所以我们可以假设在任何传统中都有一定程度的潜在创新，反之亦然。

建筑设计的发展，在其广义上讲，一直面临着这样的问题。任何时间及地点的文化，都会面对过去的传统和对未来的展望。

对于中国建筑的发展来说，20世纪末和21世纪初是试图定义传统与创新之间辩证关系的关键时期。

如果说中国传统建筑有一个共同的理念，那就是一个单层或多层的楼阁，由木柱和梁、砖墙、木框窗和倾斜的瓦屋顶构成。这一理念是从中国土地上几个世纪的建筑实践中提炼出来的，忽略不同地区建筑和不同时期之间的差异，以不同的中国封建王朝命名。

The debate around the concepts of innovation and tradition, in the many fields of human knowledge and activity, as diverse as literature and economy, as distant as agriculture and politics, is still alive, in some fields very active. Their sheer definition is uncertain, as it happens to abstract concepts encountered by different cultures, different spaces, and across several periods of time.

If we try to give them a rough meaning, abstracting from the innumerable cases of reality, we would say that tradition is the ways and the methods, coming from the past - even though an undefined historical past - with whom we are now approaching a set of problems; innovation, on the other hand, would be any unprecedented, completely new way or method using which we are facing a set of problems.

These two concepts are intertwined, and may even overlap. It has been observed by many critics that tradition is not possible without innovation, and swapping the terms the observation remains valid. Also, one innovative solution may become part of tradition, and one traditional feature may become innovative, so we can assume that in any tradition there is a certain degree of potential innovation, and vice versa.

The development of architecture design, in its most general and inclusive meaning, has always faced such questions. Any culture, in any place and time, has been confronting with the heritage from the past and with the projection into the future.

For the Chinese architectural development, the latest 20 century and the beginning of this 21 century is a crucial period for attempting a definition of the dialectic between tradition and innovation.

If there is one commonly shared idea of traditional Chinese architecture, that would consist of a single-level or multi-storey pavilion, with wooden columns and beams, brick walls, wooden framed windows and pitched tile roof. It is an idea that comes abstracting centuries of construction practice on Chinese soil, not counting the differences among regional architecture and through several time periods, named after the Chinese imperial dynasties.

But most importantly, Chinese architectural and urban development through the centuries has fixed a spatial scheme, a peculiar composition of spaces, that we may define as the courtyard house. This scheme - built

但最重要的是，几个世纪以来，中国的建筑和城市发展已经确定了一种空间规划方案，一种独特的空间构成，我们可以把它定义为四合院。这个方案围绕中心庭院建造体块，强调了对空间所起到的基本作用的论述，空并不是在一组建筑物的中心建造的空间。这个空间是四合院的中心点，通过它来分布体量并串联所有的楼阁。

这种建筑组织原则在朱锫工作室设计的寿县文化艺术中心（160页）中得到了很好的体现。该建筑看起来像一个矩形的体块，它的尺寸达到了城市建筑的规模。走近建筑，游客首先注意到的是高高的混凝土墙、平坦的屋顶，和几个比较大的洞口；然后是环绕建筑体块的水池和作为入口的低矮桥梁。

所有这些看起来都与中国传统建筑的总体概念相去甚远，但建筑师的解释揭示了建筑与当地历史的联系：巨大的立面重现了古老的城墙，高高的城墙只有不几扇门，下方环绕着护城河。

进入建筑体块，与中国传统的联系变得愈加清晰：行走于六个主庭院之中，其顶部是露天的，这里有长廊、凉廊，楼梯和小桥构成了建筑步道。庭院一层分布着一组引人入胜的树木、石头和水池。庭院的主要功能是将空气、阳光和漫射光引入建筑内部空间，将天空引入城市建筑体块。建筑本身成了老城区氛围的代表，"坚实的立面、室内庭院、连接的通道"均为当地传统建筑的主要特征。

小径、人行天桥、悬空的长廊彼此相望，是谢子龙影像博物馆的主要组成主题（182页）。在这个白色的混凝土体量上，魏春雨地方工作室的建筑师执行了许多别致的操作，如：雕刻出巨大的体块、弯曲的墙面、旋转的中心轴、悬空的桥梁，所有这一切的目的都是将参观者带入到展览的主题体验中：影像，这一最初由光影叠加在胶片上形成的产物。建筑师参考了意大利形而上画派运动，将孤立的物体和元素放置在某处，以超越其建筑实体和类型学的意义，达到"永恒与静谧"这样的绝对概念。白色混凝土的中性色彩，使建筑看似一块白色画布。

无论如何，建筑成果是显著的，当我们透过中国传统的视角来看博物馆时，我们看到的是开放庭院和封闭房间的空间通过小径和廊道以另一种方式表达了出来。所有的交通流线空间，以及几乎每个展厅，都从天窗获得自然光，在混凝土表面上创造出壮观的光影效果。沿着贯穿展览空间的环廊，我们可能会认出某些过去的设计主题：如看到门槛并跨了过

volumes around a central yard - stresses a remark on the fundamental role played by the void, empty, not-built space in the center of a set of buildings. This space is the pivot of the courtyard house, by distributing the volumes and linking all the pavilions together.

This architectural organizing principle is well evident in the Shou County Culture and Art Center (p.160) by Studio Zhu-Pei. The building appears as a rectangular block, whose dimensions reach those of an urban block. Approaching the building, the visitor first notices the high concrete walls, the flat roof, few big openings; then a water pool surrounding the block, and low bridges as entrances.

All of it may appear very distant from the general idea of a Chinese traditional building, but the architect's explanation reveals the link with the local history: the massive facades reproduce the ancient city walls, full height walls with few gates, circled by water moats.

Entering the block, the link with Chinese tradition becomes clearer: we walk through six main courtyards, always opened up to the sky, with galleries and loggias, stairs and bridges as architectural promenade, and on the yards' ground floor a scattered but evocative set of trees, stones and pools. The yards, of course, have the main function of bringing air, sunlight and diffused light to the block's inner space, taking the sky into the urban block. The block itself becomes a representation of the old city atmosphere, where "solid facades, interior courtyards, pathways connecting" are the main features of the local traditional architecture.

Pathways, footbridges and suspended galleries overlooking each other, are the main composition theme of the Xie Zilong Photography Museum (p.182). On this white concrete block, the architect WCY Regional Studio has carried many different operations, like carving out the massive block, bending wall surfaces, rotating axis, hanging bridges, with the goal of involving the visitor into the exhibition theme experience: photography, which is primarily made of dark and light impressed on film. The architect refers to the Italian Metaphysical art movement, where solitary objects and elements are positioned in order to transcend their physical and typological meaning for reaching such absolute ideas like "eternity, tranquility", written on the neutrality of white concrete, like a white canvas.

©Schran Images

寿县文化艺术中心
Shou County Culture and Art Center

©Yao Li

谢子龙影像博物馆
Xie Zilong Photography Museum

去。虽然在每个中式庭院中，穿过不同房间的通道总是以木台阶为标志，但在博物馆中，它依赖于明暗对比。

水池围绕在建筑四周，外墙和天空倒映在水池中，弯曲的人行道被称为"交流之路"，强调了步道的体验所具有的核心作用。这座建筑乍一看上去如同一个坚不可破的堡垒。

对于阿兰若艺术中心的设计（218页），Neri & Hu的建筑师们以另一种方式探讨了同样的主题，即新建筑与传统的关系。该项目仍然是一个封闭的类似立方体的体量，虽然其形状仅仅是为了适应城市环境，但保持所有的立面均为纯粹的矩形，只有很少的大型门窗，以及一种较小的洞口，由重复的模块制成。室内围绕着一个中心空间组成，中央有一个向天空开放的倒立的圆锥体，将光线和空气引入各个楼层的长廊。在一个相对包容的体量中，外缘为方形，露天核心为圆形，这两种不同的几何形状之间的鲜明对比是解读整个设计概念的关键。

尽管中央锥体不是连接几个长廊房间的实体通道，但却是一个看得见、摸得着的交流元素：圆锥空间在每一层的显著位置，通过将主要的交通流线设计为弧形，墙壁和房间内部的空间也随之确定。圆形的几何形状揭示了中心连接空间的重要性。

因此，建筑展示了中心空间的传统主题，那就是"空"，它将天空带入建筑体量，并连接了周围的所有房间。建筑师将中央的圆锥形空间称为"公共空间，用于集水，吸收自然光"。中心空间的最底部是一层公共空间，那里是一个用于表演的小礼堂，也是一个极具特色的空间。

The result is anyway remarkable, but when we look at the museum through the Chinese tradition glasses, we see that it is another expression of the spatial concept of open yards and enclosed rooms linked by pathways and passages. All circulation spaces, and almost every exhibition room, receive natural light from skylights, creating spectacular light-shade effects on the concrete surfaces. Following the circulation loop across the exhibition spaces, we may recognize other design themes of the past, like the threshold, and the act of trespassing it. While in every Chinese courtyard-house the passage through different rooms was always marked by a wood step, on the museum it relies on light and shade contrast.

Surrounding the building, the presence of water pools reflecting the walls and the sky, and the tracing of curved pathways, called "communicating path", highlight the central role played by the experience of passage along, by and through a building that appears, at first sight, as an impenetrable fortress.

With the Aranya Art Center design (p.218), the architects Neri & Hu approached the same subject, the relation of a new building with tradition, in yet another way. The block is again a hermetically enclosed cubic-like volume, shaped only in plan to adapt to the urban situation, but keeping all facades as pure rectangles, with few large doors and windows, and a pattern of smaller openings, framed by a repeated module. The interior space is instead composed around a central void, an inverted cone opened to the sky, that carries light and air to all the galleries floors. The striking contrast between the two different geometries, squared on the outer edge and circular for the open-air core, in a volume of relatively contained dimensions, is the key for reading the whole design concept.

Even though the central cone is not a physical place for passing from one gallery room to the next, it is a visual and tangible element of communication between them: the conic void is well evident at any floor, by curving the main circulation pathways and thus shaping the walls and the rooms' space. The circular geometry reveals the importance of the central connecting space.

The building thus displays the traditional subject of a central space, empty, that brings the sky into the built

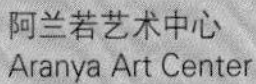
阿兰若艺术中心
Aranya Art Center

长江美术馆
Changjiang Art Museum

该项目展示了四合院的传统空间设计，即使应用于有限的建筑表面积和多层建筑，依旧保持着其价值的完整性和有效性。

由直向建筑设计事务所设计的长江美术馆（200页）也是一个立方体量，在角落处理上展示了一些突破传统的解决方案，立面全部覆盖了特殊定制的砖，并展示了带有悬梯、悬挑雨篷和人行桥的混凝土结构。该建筑体块嵌入太原市的一个新开发区，这里曾经是市郊的乡村，如今周边都变成了高层住宅大楼，因此它的建筑理念是构建"当代对过往碎片记忆的回应"。建筑师的这句描述让我们关注到了当下与那些忘却的记忆是相连的，它们曾经真实存在于那里，应被重新拾起。我们可以将它理解为将建筑与四合院的空间布局联系在一起的纽带：一个被建筑体量包围的中心空间，一组交叉的步道将其连接在一起。

建筑的主要公共通道完全独立于馆内的展览路线，从位于街面的主楼梯开始向上爬升，通向二层的露台，这个露台是一个"被抬升的广场"，广场中央庭院种有树木。通道一直通往北侧的步行天桥。其开放的室外散步空间将建筑体量所占据的空间还给了公众，丰富了建筑的特色。

博物馆内部的交通流线也从一层开始，位于南角的一个圆形拱顶是连接各空间的核心，围绕它有一部螺旋楼梯，由下至上连接所有的展馆。画廊和圆形空间的采光均来自天窗，在顶层的设计是将均匀的自然光引入房间。在非露天的地方，穿孔的屋顶将过滤的阳光引入室内。

volume, and connects all the rooms around. The architect refers to the central conic void as a "communal space, for water collection, and bringing light from the sky". On ground floor, the central void ends in the communal space, a small auditorium for performances, a space of extraordinary character.
This project demonstrates how the traditional spatial scheme of the courtyard house, even when applied to a limited surface and multi-storey building, maintains its values intact and efficient.
Changjiang Art Museum (p.200) by Vector Architects is also a cubic volume, displaying several heterodox solutions at the corners, facades fully covered in a specially designed brick, and shows up the concrete structure with hanged stairways, cantilevered canopies and a footbridge. The block is inserted in a new development area of Taiyuan, surrounded by residential towers built in place of a former suburban village, so it was conceived as the "contemporary response to a fractured memory". This statement by the architect brings to our attention the link to a lost memory, something that lies in the past and should be taken back. We may read it as the link that binds this building to the spatial scheme of the courtyard house: a central void surrounded by built volumes, and a set of pathways crossing and tying them together.
The main public path, independent from the museum inner circulation, takes start from an open stairway at street level, it leads to the terrace at second level, a "raised up plaza" with trees, and ends up on the footbridge to the northern side; this pathway gives back to the public what was taken by the built volumes, in terms of open air promenade-space, enriched by the architectural features of the building.
The museum inner circulation also starts from ground floor, but the main connecting core is a circular vaulted space in the southern corner, around which a spiral stair raises up, linking all the exhibition galleries. The galleries and the circular void are both enlightened from skylights, which at the top floor are designed to spread over an equal natural light into the rooms. Where the sky is not at open-air sight, there comes the perforated roof to bring the filtered sunlight in.
The particular attention paid by the architect to circulation pathways and natural lighting, strengthened by

建筑师对交通流线通道和自然采光的特别关注，着重体现在采用了凸出的窗户、旋转的砖砌百叶窗、悬空的玻璃盒子结构和方形的混凝土天花板上，这使得这个实体体块与周边的城市环境比起来，显得格外与众不同，引人注目。

然后，这些项目可以被解释为中国传统建筑的一个独特特征的延续，特别是四合院的空间方案，因为它们都被构思为一个中央空间或主要的空间，围绕着它，主要的交通流线通道连接了所有建成的体量，或是封闭的房间。所有这些建筑至少都秉承这种建筑理念，和其他传统特征比起来，这一特点尤为明显。中国的传统跨越千年，我们总能找到这样一些令人回忆过去的建筑元素作为借鉴：那些花园庭院中的水池、树木、植物；高高的城墙、光影浮动的窗框、圆形的洞口以及串联庭院和建筑的小路。

另一方面，这些项目在哪些方面体现了创新性？为了了解它们的创新之处，我们可以关注中国建筑史的最新发展：看看历史的传承，我们就可以发现当代项目的新颖性。

近20年，中国的建筑走向多元，从公共建筑、基础设施到私人住宅和室内设计，在各个领域都取得了令人瞩目的成果。近40年来，国家在政治和经济上的投入迅速改变了中国的面貌，使得建筑研究路径的发展成为可能。中国从一个以农村为主的国家，转变为一个现代的、城市化的、具有合作性的世界大国。

许多建筑设计之路一直在努力寻找新的"中国现代性"的定义，将识别"中国风格"的需要与对当代技术和方法的运用相结合。结果大多数建筑都被设计为具有夸张的尺寸，彰显财富、展示规模和工业实力、强化象征主义，让人不禁想起中国建筑风格的理想化版本。

除了这些主流建筑之外，一代通晓近现代建筑知识的中国建筑师还探索了其他的路径。他们往往拥有国外学习和生活经历，除了了解中国千年的建筑传统之外，更深谙当地和中国的古老传统，包括其他艺术领域，如景观和园林设计、素描、油画和雕塑。

protruding windows, rotated brick shades, a hanged glass box and the concrete squared ceilings, is what makes this solid block different and outstanding compared to its urban context.

These projects may be then interpreted as continuation of a particular feature in architectural Chinese tradition, specifically the courtyard house spatial scheme, as all of them have been conceived with a central or dominant void space, around which the main circulation pathway connects all the built volumes, or enclosed rooms. All of them share at least this architectural concept, especially recognizable among other "traditional" characters. In the millenary Chinese tradition we may find the references that these projects intend to recall and bring back: water pools, trees, plants and bridges as in garden design, high city walls, window frames filtering light and shadow, circular openings, pathways passing through and linking different yards and buildings.

In what ways, on the other hand, the projects display innovation? To understand where lies their innovative character, it is convenient to focus at the recent developments of architecture history in China: looking at the past heritage we may find the novelty of the contemporary projects.

In the latest 20 years, architecture in China has taken many different directions, showing impressive results in every field, from public building and infrastructures to private housing and interiors. The development of these architectural research paths was made possible by political and economical inputs that changed rapidly the face of China in the latest 40 years, from a mainly rural country to a contemporary, urbanized, interactive global power.

Many of the architectural design paths have struggled to find a definition of the new "Chinese modernity", mixing up the need of identifying a "Chinese style" with the use of contemporary techniques and methodologies. Most of the outcomes still end up in buildings of exaggerated dimensions, displaying wealth, a showcase of bigness and industrial power, and enhanced symbolism recalling an idealized version of the Chinese architecture style.

Aside this mainstream production, some other paths were traced by a generation of Chinese architects who gathered knowledge on modern and contemporary architecture, often by abroad studies and experiences, and enriched it with knowledge of the local and ancient Chinese tradition, including other artistic fields, like landscape and garden design, drawing, painting and sculpture, beside the millenary building tradition.

入选的项目就是后一种研究路径的结果，因为它们巧妙地将建筑材料的现代性和具有丰富当地传统的当代建筑语言相结合，避免了空有闪亮外壳的“中国风格”的形式呈现。

他们的创新在于对材料的熟练使用，比如，混凝土结构和墙壁、钢结构和覆层，甚至有几百年历史的砖墙、宽幅玻璃表面等等，利用每一种材料的潜力，创造出一种在中国前所未有的建筑语言。我们可以在这些项目中看到的具有创造性的语言，包括悬索桥、天篷和悬挑的楼梯、箱型结构、悬臂墙；平屋顶、屋顶天窗、工业生产的立面模块；就像在一个巨大的体量上进行雕刻、折叠或扭转的操作。

虽然这种翻新的建筑语汇可以用于恢复一些过去的空间特征，但它显然不属于中国传统。但正如开头所提到的，创新和传统是相互交织的，也可能是重叠的：当我们现在看到这种由混凝土、钢铁或玻璃构成的建筑语言时，它可能会在未来成为一种新的“传统”设计方法。

此外，创新也可以在城市层面上找到，因为这些建筑都扮演了新的城市角色。它们的公共空间对游客和路人开放，被博物馆的展品隔开，这个空间不局限于外部花园和通道，但是要通过建筑体块进入。这种非常现代化的设计方法可以用来实现建筑物的公共功能和优化公共地面的使用。通常，城市和郊区的公共空间局限于人行步道和封闭式公园，因此，了解创新型城市公共建筑的特点就显得尤为重要，就如同本书提到的博物馆建筑。

因此，创新明显体现在对于现代材料的熟练使用、由此而来的更新的建筑语言，以及这些公共建筑在城市中的作用上。

最后，创新还体现在对当地传统的高度关注上，认真考量历史遗产的形式和空间价值，尊重中国的文化认同，并借助新工具、新方法和新的设计语言对其加以阐释。

归根结底，创新就在于它让传统的四合院空间布局再一次走进今天，走向未来。

The selected projects are a result of the latter research path, as they skillfully combine the modernity of materials and the contemporary architectural language with the richness of the local tradition, avoiding any formal representation of the "Chinese style" in just a new shiny vest.

Their innovation lies then in the mastered use of materials, like concrete structures and walls, steel structures and cladding, even centuries-old brick walls, wide glass surfaces and so on, taking advantage of each material's potential in creating an unprecedented, in China, architectural language. The innovative language we can see in these projects consists of suspended bridges, canopies and overhanging stairways, boxes, cantilevered walls; flat roofs, rooftop skylights, industrially produced modules for facades; figurative operations like carving, folding or twisting a massive block.

Although this renovated architectural vocabulary may be used to recover some spatial features coming from the past, it doesn't belong, evidently, to the Chinese tradition. But as mentioned in the beginning, innovation and tradition are intertwined and may overlap: as we now look at this language made of concrete, steel or glass, it may become in future times a new "traditional" design method.

Furthermore, innovation can be found on urban level, for the new urban role that these buildings all play. Their public space, open to visitors and simple passers-by, and separated by the museum exhibition, is not limited to the outer gardens and pathways, but it enters through the block. This very modern approach to the public function of the building and the use of its public ground, in cities and suburbs where often the public space is limited to sidewalks or enclosed parks, is an important key to read the innovative urban character of a public building, like a museum is.

Innovation, thus, is well evident for the mastered use of modern materials, for the consequently renewed architectural language, and for the urban role of these public buildings.

Finally, innovation lies as well in paying special attention to the local tradition, in taking in high consideration the formal and spatial values of the consolidated heritage of the past, the Chinese cultural identity, and translating it with renovated tools, new methods, and fresh architectural language.

It resides, in the end, in taking exactly that traditional courtyard house spatial setting, and transporting it once more into the present and future times.

寿县文化艺术中心
Shou County Culture and Art Center
Studio Zhu-Pei

在寿县建造的一座封闭的矩形体块，借鉴了当地传统民居的隐藏式庭院风格

An enclosed rectangular block in Shou County references traditional local house hiding courtyards

安徽省寿县位于淮河南岸，历史上曾是楚文化的故乡，也是淮南王刘安著书立说的地方。古城平面略呈方形，围以夯实的、稍微倾斜的土墙。墙体以土夯筑，外侧砌砖，其下方是一个抬高的石头地基。城外东南为护城河，北环淝水，西接寿西湖，远眺八公山，漫步在古桥之上，漫步在老城的古迹中，都能给你带来无限的灵感。城墙将城内凌乱分散的建筑包裹得紧紧的，在跨入城门的那一刻，顿时会有突变之感。

这座城市的庭院建筑既不同于北方的院落住宅，也有别于安徽南部的徽州民居。这些建筑类型强烈地反映了这个区域的自然气候特征，以及当地的生活方式。沿着狭窄的街道看去，垂直的四合院建筑通常有小窗户和坚固的墙壁，以抵御风雨。

寿县文化艺术中心建在古城的东南一两公里外的一个地块上，过去是农田，空旷、平坦，毫无特点。多数新建筑既不反映当地的自然气候特征，又与当地的文化没有一点关联。苍白的地段却为朱锫提供了相当大的创作潜力。县政府没有就设计提具体要求，但时间迫在眉睫。

文化艺术中心需要包含美术馆、文化馆、图书馆及档案馆等功能。在中国，这类文化艺术中心并不算是新鲜事物，受影响于苏联，在20世纪50年代到70年代建造了很多。但今天，它又被赋予很多新的可能，包括为市民体验文化活动提供城市公共空间。

建筑师从寿县的文化根源中获取灵感，观察了古老的民居和废墟，感悟我们的先人在没有现代技术的前提下，是如何在原始自然力量和建造中寻求平衡的。垂直四合院住宅的内向型家居形态、四通八达的狭窄里弄将家家户户彼此相连，这些特征不仅映射出当地人的生存方式，也暗示出当地自然气候的建造法则。建筑师朱锫重新构想了这样一种空间经验，让寿县文化艺术中心根植于当地的社区之中。

大小不同的多个庭院被置于一个封闭的矩形体块中，一条起伏的公共廊道将它们串联起来。这条漫游环道引导着人们缓步桥上，穿越壕沟步入建筑之中。建筑主入口有一个广阔的前院，形成了一个公共广场，代表了典型寿县民居中的"堂屋"（中间的房间），而后院则类似于当地的民居。该建筑的每个功能区都有两三个内庭院。从前院开始，访客可以穿梭游览于布满了很多内院的整座建筑，完全不会有间断感。沿着这条可遮阳避雨的环形廊道，访客时而在一层，时而又到了二、三层，空间变幻莫测，光影时明时暗，去感悟"藏、息、修、游"所赋予中国传统建筑的艺术精神。

寿县文化艺术中心的内向型概念兼顾了对周边未来发展不可预测的考量，也映射了寿县古城历经时代变迁依旧泰然处之、兼容并蓄的强大的生命力。

Shou County, Anhui Province, is located on the south bank of Huai River, the historical center of the Chu culture, and the place where Liu An, King of Huainan, edited his compendium of ancient Chinese philosophy and poetry. The old town is roughly square shaped, surrounded by rammed, slightly inclined, earth walls, with the outward-facing sides covered in brick above an elevated stone foundation. A moat lies to the southeast, the Fei River to the north, Shouxi Lake to the west, and the Bagong Mountains in the distance. A walk across the ancient bridges and a stroll through the ancient relics of the old town can offer boundless inspiration. The city walls wrap tightly around chaotic and scattered buildings, creating a sudden change in experience on crossing the threshold.

The city's courtyard buildings differ from those in northern China and southern Anhui. The typology strongly reflects regional climatic conditions, as well as the local way of life. Along narrow streets, vertical courtyard houses typically have small windows and solid walls for weatherproofing.

Shou County Culture and Art Center was built one or two kilometers southeast of the old town, on flat, featureless farmland. Most new buildings there are generic and fail to reflect the

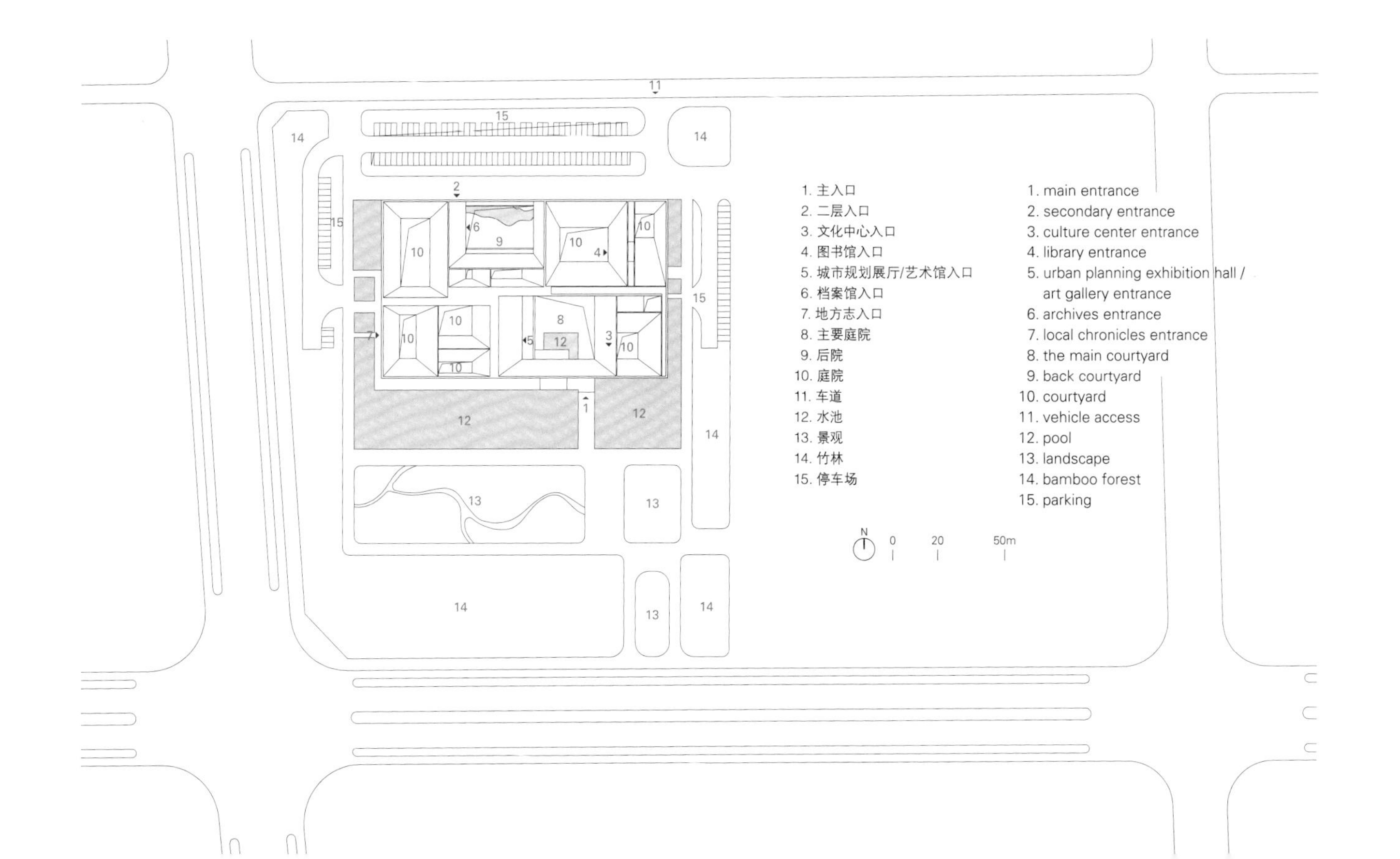

安徽民居类型
local Anhui housing typology

实体立面 solid facade

内院 interior courtyards

过道连接庭院
pathways connect courtyards

概念图
concept image

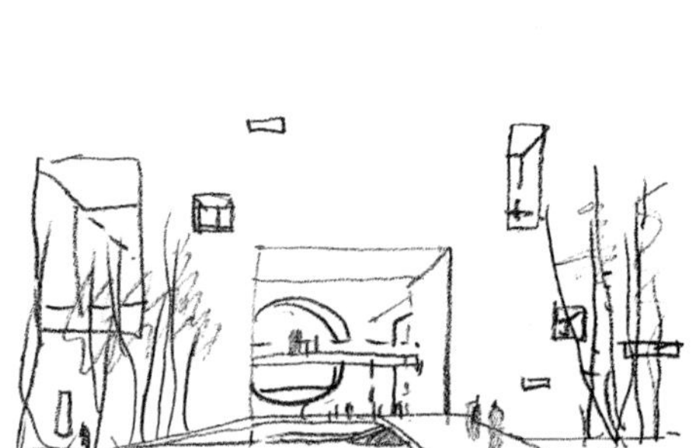

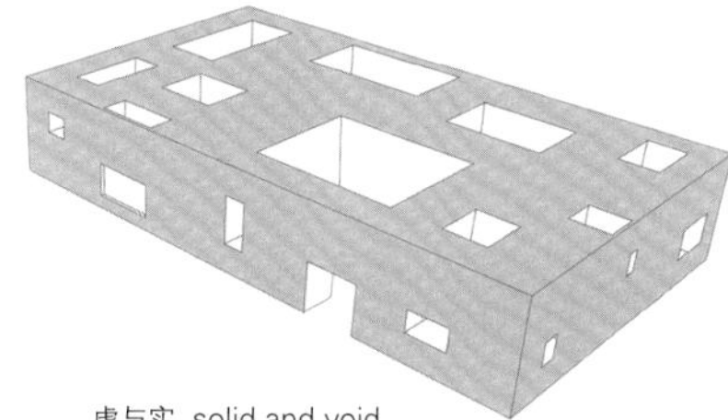

虚与实 solid and void

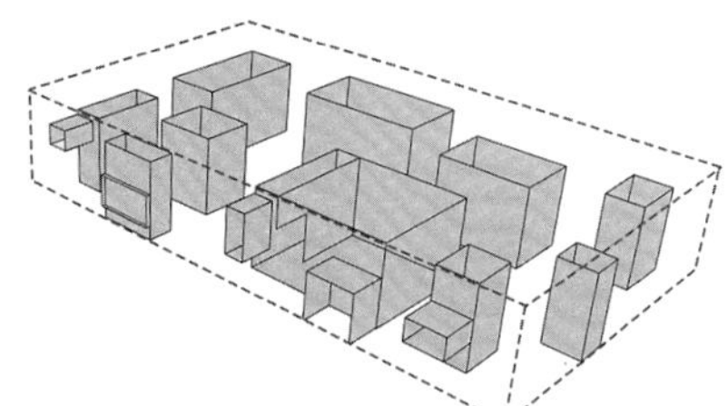
内院 inner courtyards

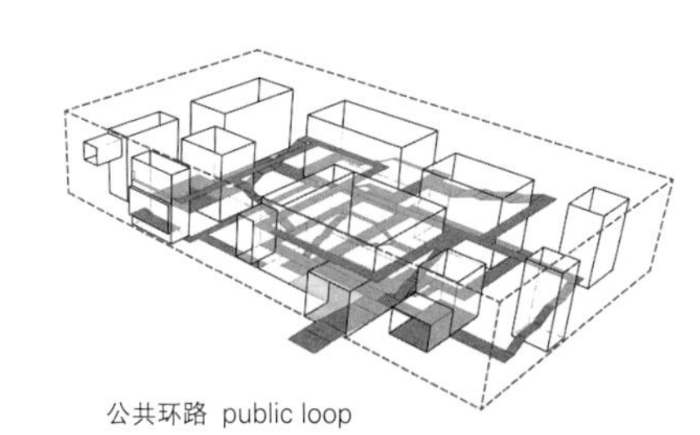
公共环路 public loop

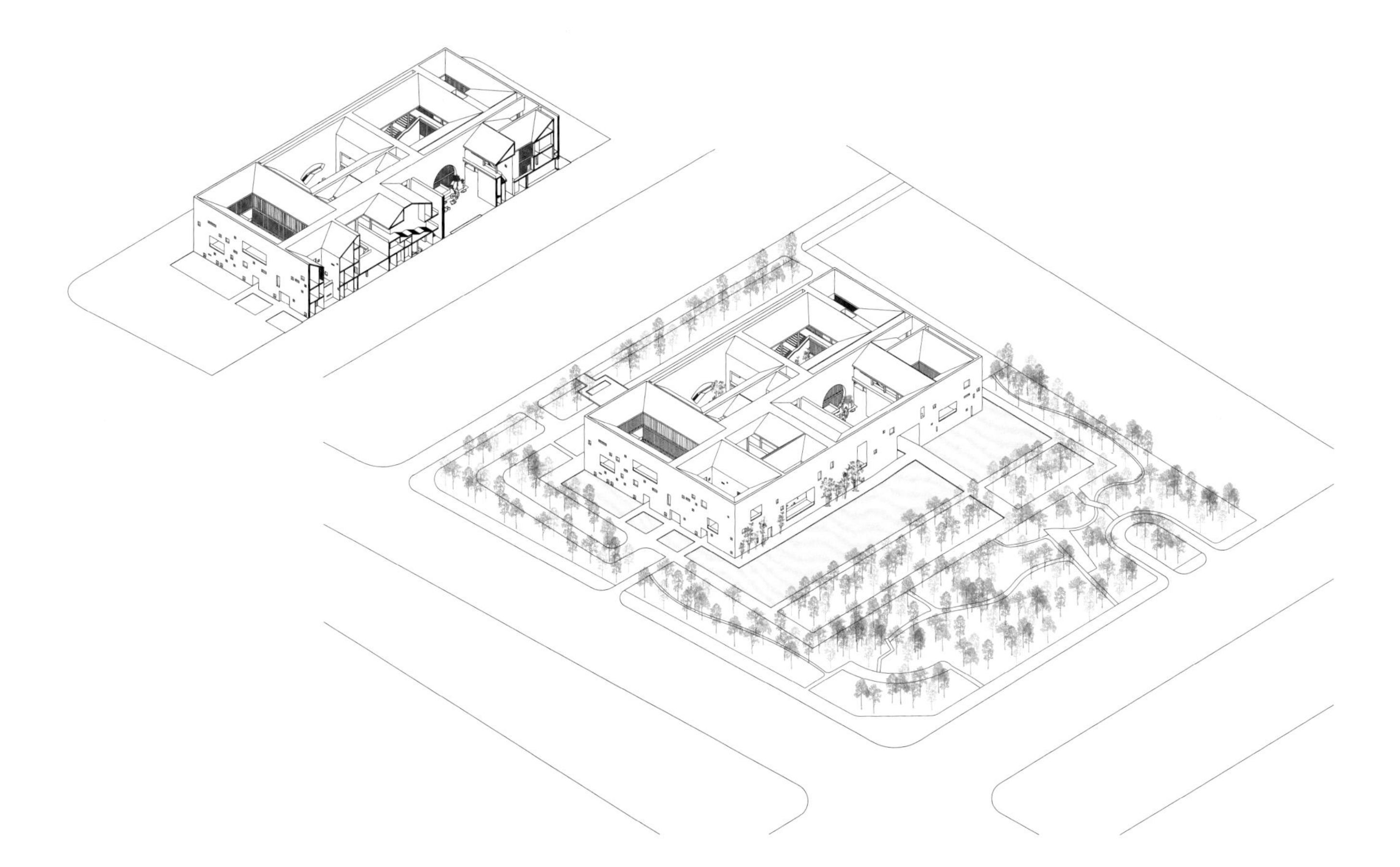

项目名称：Shou County Culture and Art Center / 地点：Shou County, Anhui, China / 建筑师：Studio Zhu-Pei / 设计主管：Zhu Pei / 设计团队：You Changchen, Shuhei Nakamura, Du Yang, Liu Ling, Wu Zhigang, Yang Shengchen, Ding Xinyue, Ke Jun, Wu Zhenhe, Duyao / 总承包商：ShengWo Construction Group Co., LTD. 结构与机电顾问：BIAD JAMA CO., LTD. / 景观与室内设计顾问：Studio Zhu-Pei, The Design Institute of Landscape & Architecture China Academy of Art / 供货商：concrete-Anhui Huacheng Concrete Co.; aluminum curtain-Shandong Jinxiang Aluminum Co.; glass-Wuxi Yaopi Glass Engineering Co. / 客户：Shouxian Government / 总面积：30,010m² 竣工时间：2019 / 摄影师：©Schran Images (courtesy of the architect)

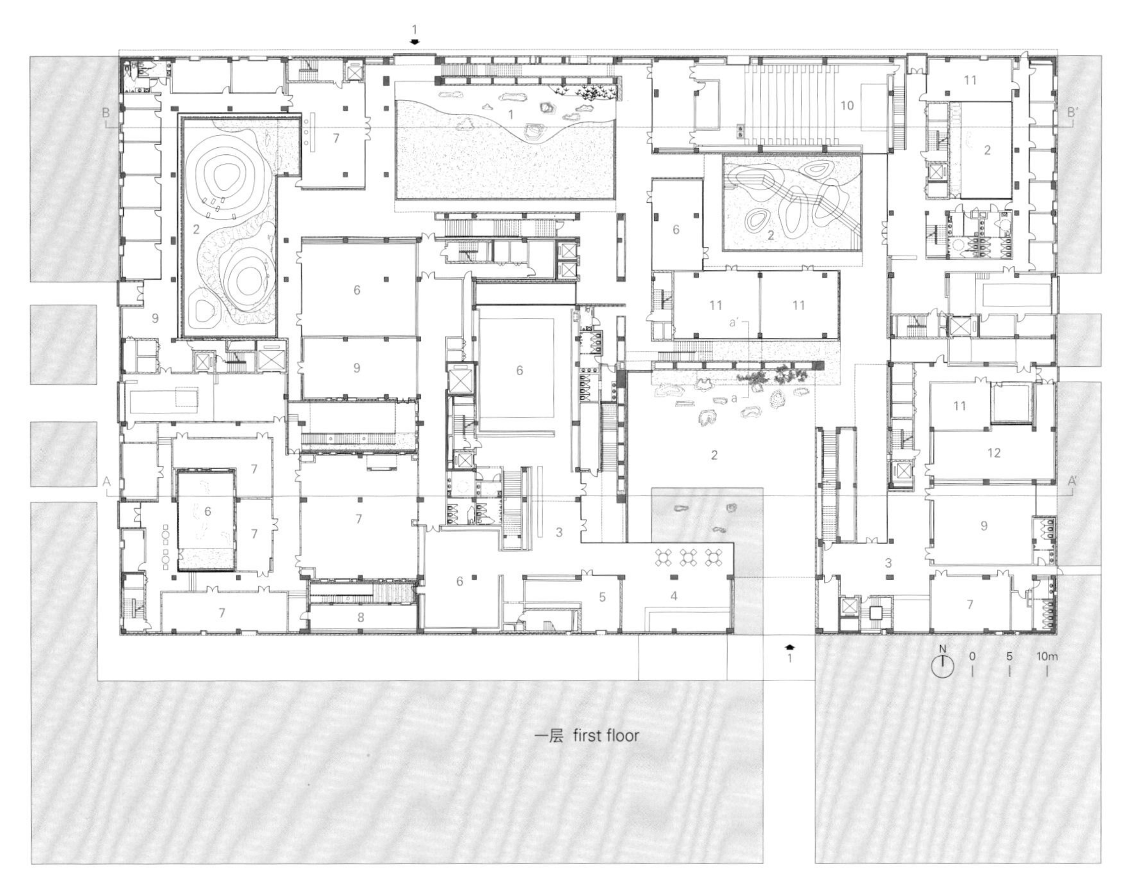

一层 first floor

1. 主入口 2. 庭院 3. 接待处 4. 咖啡厅 5. 书店 6. 展览空间 7. 档案馆 8. 储藏室 9. 教室
10. 阶梯教室 11. 阅览室 12. 多功能室 13. 多媒体室 14. 活动室 15. 工作室 16. 设备间

1. main entrance 2. courtyard 3. reception 4. cafe 5. book shop 6. exhibition space 7. archives 8. storage 9. classroom
10. lecture room 11. reading room 12. multi-function room 13. multimedia room 14. activity room 15. studio 16. equipment room

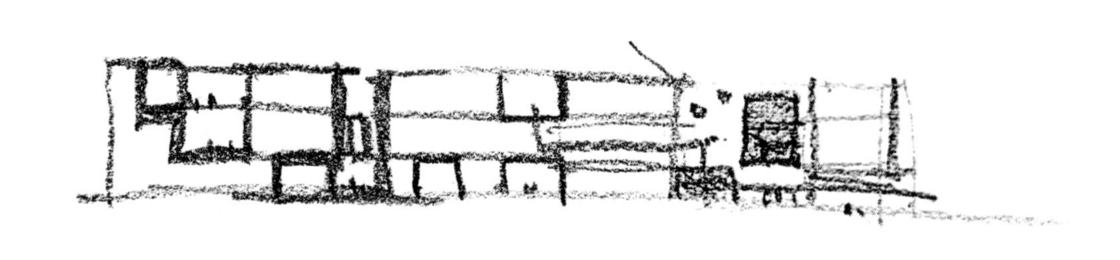

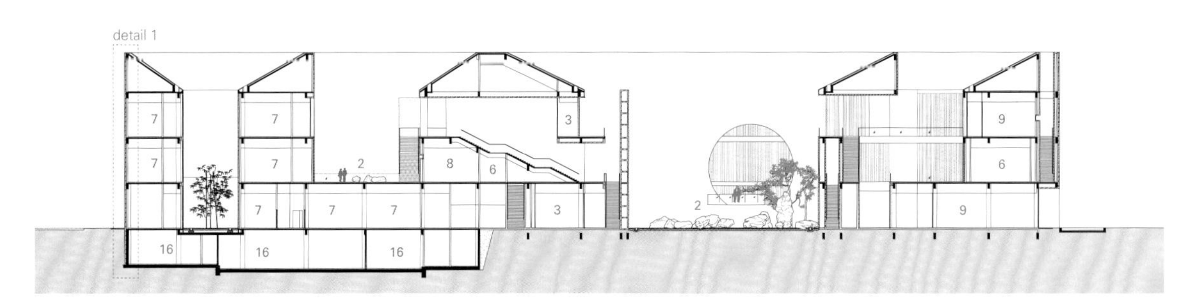

A–A' 剖面图 section A-A'

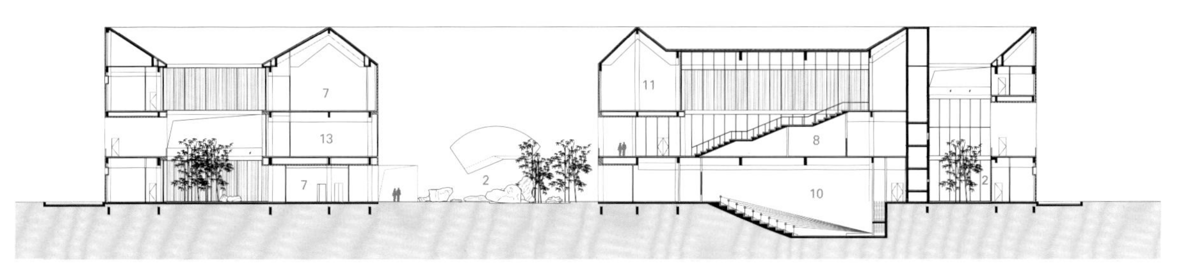

B–B' 剖面图 section B-B'

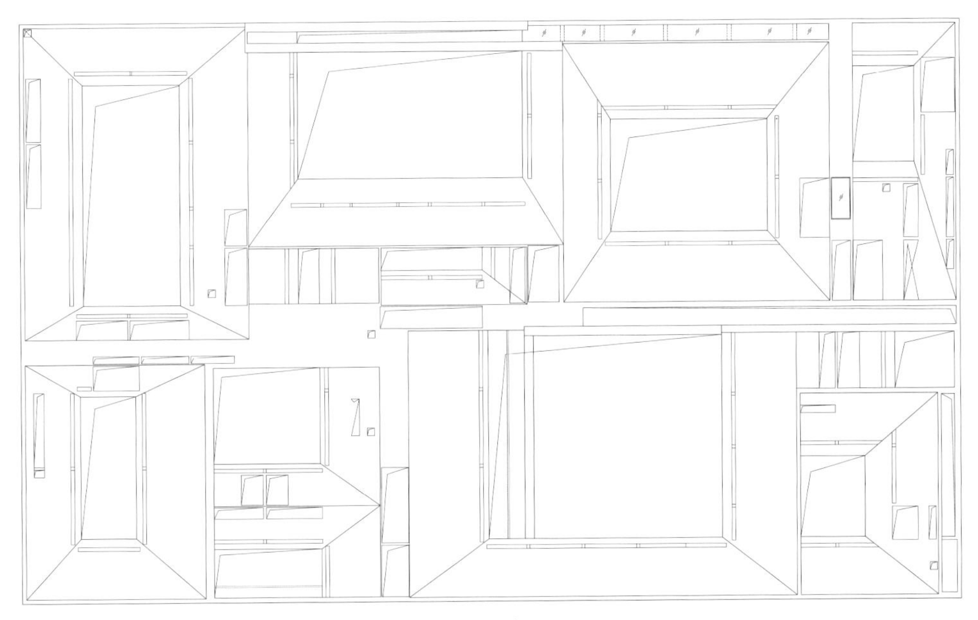

屋顶 roof

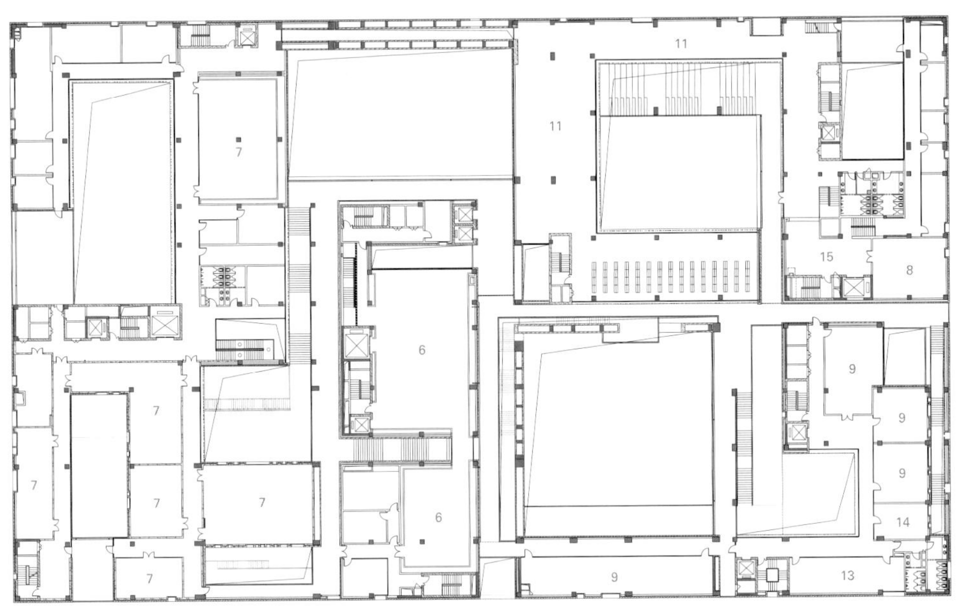

三层 third floor

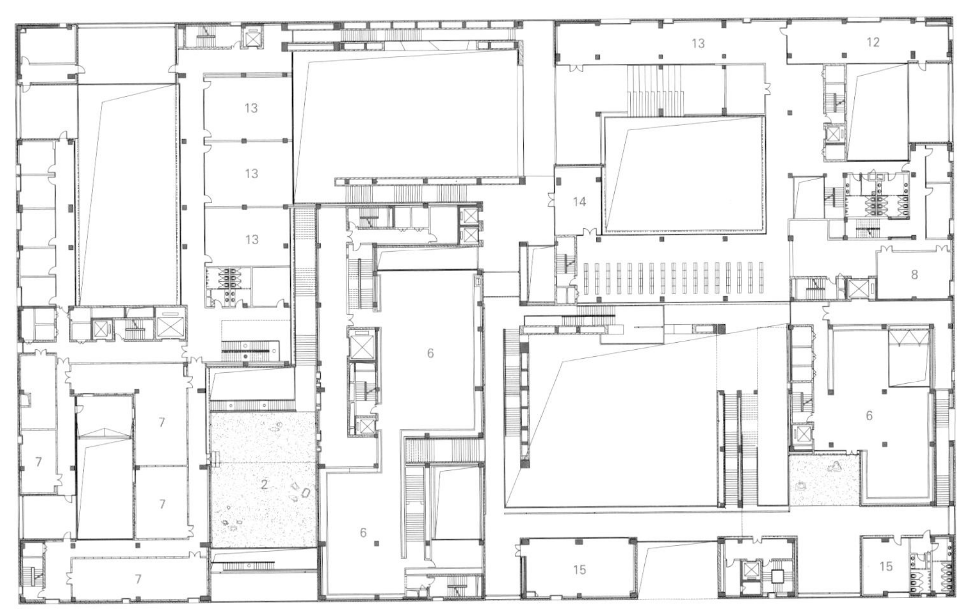

二层 second floor

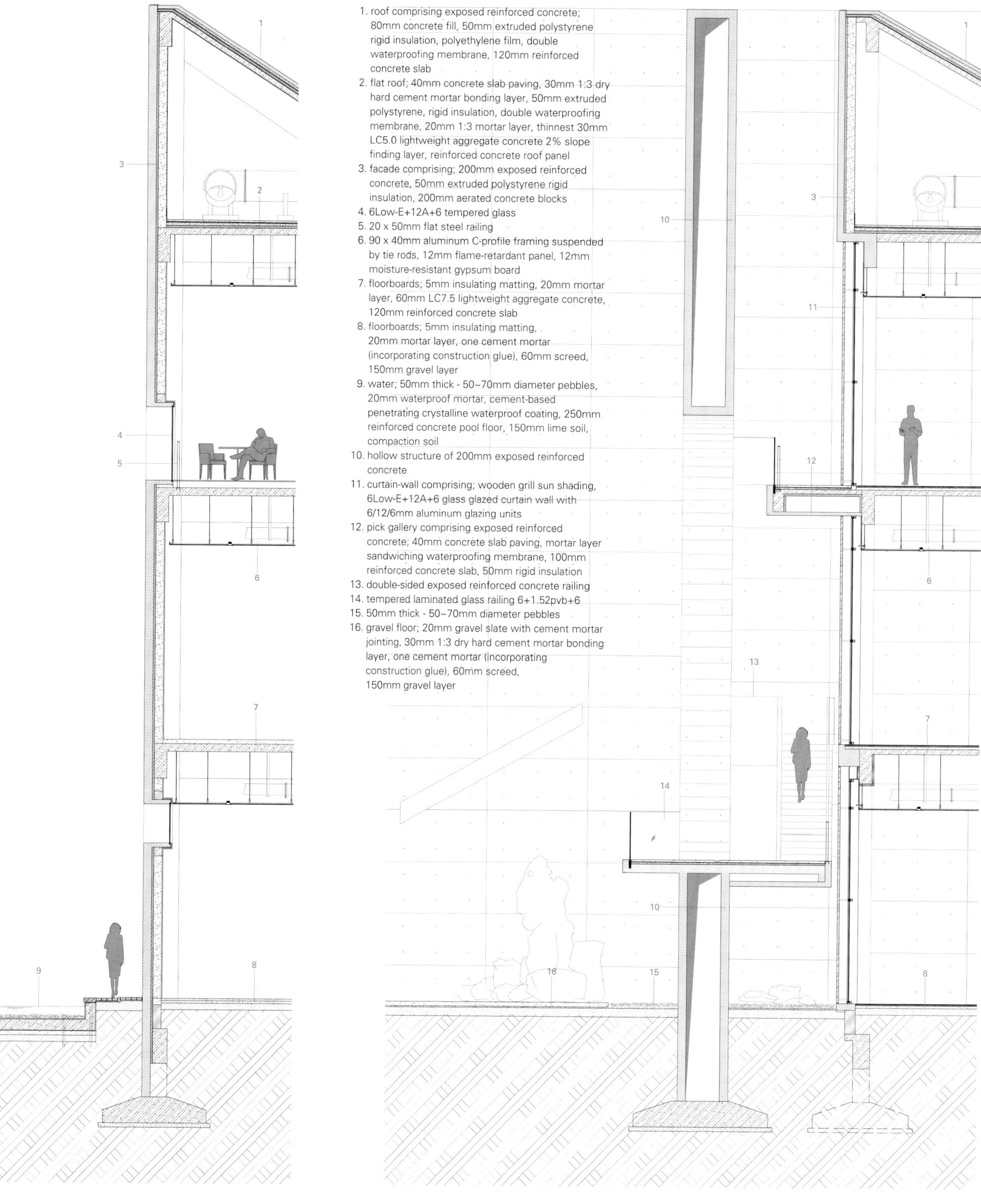

详图1 detail 1

a-a' 剖面详图 detail a-a'

local climate and culture. The blandness of the project site, however, offered Zhu Pei considerable creative potential; the county government did not insist on specific design requirements, but time constraints were tight.
The building needed to include an art gallery, cultural center, library, and archive. Such centers are not new in China – many being built between the 1950s and 1970s with Soviet influences – however today they present more possibilities, including creating urban public space for cultural activities to be experienced.
The architects drew inspiration from the cultural roots of Shou County, observing old dwellings and ruins, and identifying how ancestors balanced the primal force of nature and construction without modern technologies. The inward-oriented living patterns of the vertical courtyard houses, and the narrow lanes connecting them, reflect the local way of life and hint at the rules of construction for local climate conditions. Zhu Pei reimagined this spatial experience to help to embed the art center in the local community.

Multiple courtyards of different sizes are placed in an enclosed rectangular block, connected by an undulating public walkway. This loop guides people over the bridge, crossing the moat into the building. The extensive front yard at the main entrance forms a public square that represents the tang wu (central room) of typical residences, while the back yard resembles those of local folk houses. Each program of the building has two or three inner courtyards. From the front yard, visitors can wander in all courtyards without interrupting the continuity of the rooms. Walking along the loop, visitors find themselves at times on the first floor, second floor or third floor. The unpredictable space, with light and shadow continually shifting, allows visitors to feel the artistic spirit of traditional Chinese architecture expressed by the principles of "hide, breathe, cultivate, and wander".
The introverted concept of Shou County Culture and Art Center takes the unpredictable future development of the surrounding area into account. It also reflects the grace, inclusiveness, and vitality of an ancient city, which has withstood the test of time.

谢子龙影像博物馆
Xie Zilong Photography Museum

WCY Regional Studio

谢子龙影像博物馆

Xie Zilong photography museum suggests to the collective unconsciousness about nature and memory

谢子龙影像博物馆的设计从思考建筑、城市与自然的关系开始。该建筑位于现有的生态湿地上，揭示了建筑与城市之间的联系，从不同的角度重新审视了城市景观建筑：建筑、人、自然是平等的，可以共生共存，拥抱世界。建筑本身是一个巨大的"艺术独立体"；施工只使用白色清水混凝土和高度密集的现场集成铸造方法。紧凑的结构被建筑师称为"凝聚的建筑诗歌"——设计考量并回应了"节俭崇拜"的传统思想。

影像博物馆是一个储存记忆和唤醒观众反思的地方。一系列的构成要素，如十字路口、拱桥、木板路、小码头等，暗示了人们对自然和记忆的集体无意识。博物馆利用日常生活中熟悉的物品，营造出一个风格迥异却又好似时空停滞的迷宫，引导观赏者关注体验的陌生感和物质性，也被称为异化体验或疏离体验。

建筑主体是一次性现浇混凝土。为了追求清水混凝土的质感和性能，在施工初期进行了多次试验。样板墙及白色混凝土的级配试验采用优质白水泥代替传统硅酸盐水泥作为主要的胶凝材料，精选白色或浅色砂石为骨料，同时剔除粉煤灰和矿粉等对混凝土白度影响较大的活性矿物掺合料（如粉煤灰和矿粉）。这一系列的改善试验使混凝土达到了预期的效果。水泥白度超过88%的国际标准。它是中国最白的清水混凝土建筑。

近年来，在政府的强有力的领导下，中国发展迅速。与此同时，涌现了一大批具有纪念意义、高投资、大规模的政治地标建筑。多数建筑貌似英勇地进入城市，却不顾城市背景或人文关怀，甚至许多建筑都缺乏细节创新。城市文化建设迫切需要民间资本的支持。中国呼吁建筑具有创新的细节、高质量和完整性。

作为湖南省第一家由私人资本全资打造的博物馆，从设计到最终投入使用，这家博物馆已成为艺术标杆。它将提升城市文化标准，改变公众对建筑的心理认知，颠覆了曾一度深深地植根于市民理念中的公共建筑被赋予的强烈政治象征。它已成为"城市客厅"、儿童艺术启蒙基地和当地艺术家的孵化摇篮，使艺术回归生活，回归大众。

The design of Xie Zilong photography museum begins by thinking about the relationship between architecture, city and nature. The building, on an existing ecological wetland, reveals the connection between architecture and the city, and provides a re-examination from a different perspective on urban landscape architecture: architecture, people, and nature are equal, and can grow together and coexist to embrace the world. The building itself is a huge "artistically self-contained body"; the construction uses only white fair-faced concrete and a highly intensive, on-site integrated casting method. The compact structure is referred to by the architects as "condensed architectural poetry" – the design thinks about and responds to the traditional idea of "thrift-worship".

The photography museum is a place to store memories and awaken viewers' reflections. A series of constituent elements such as cross paths, arc bridges, boardwalks, and small docks suggest to the collective unconsciousness about nature and memory. The museum uses a familiar object in daily life to build a maze full of differences yet also stagnated time, guiding the viewer while calling them to pay attention to the strangeness and physicality of the experience – described as alienation or estrangement.

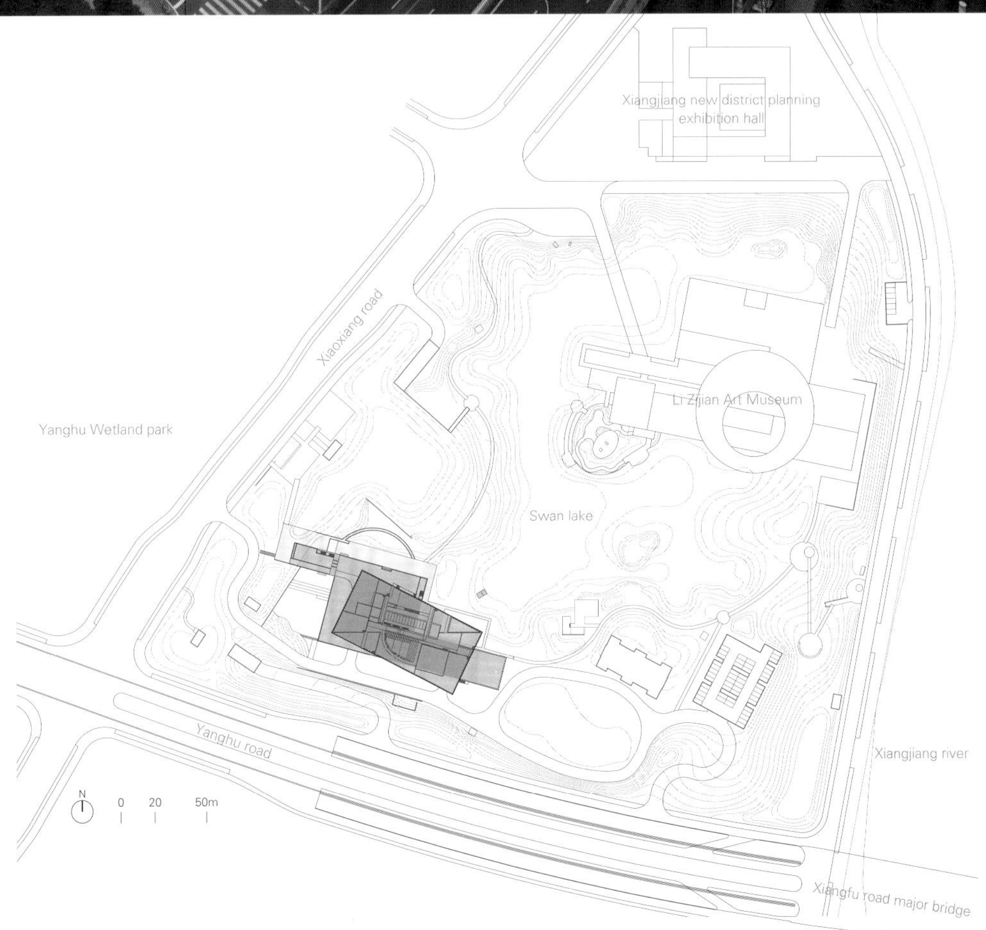
Xiangjiang new district planning exhibition hall
Xiaoxiang road
Yanghu Wetland park
Li Zijian Art Museum
Swan lake
Yanghu road
Xiangjiang river
Xiangfu road major bridge
N
0
20
50m

The main body of the building is one-off cast of in-situ concrete. In order to pursue the quality and performance of fair-faced concrete, several experiments were made in the early stages of construction. A sample wall and white concrete grading test resulted in the use of high quality white cement instead of traditional Portland cement as the main gelled material, and the selection of white or light colored sand as aggregator. At the same time, the active mineral admixture (such as fly ash and mineral powder), which has great influence on the whiteness of concrete, was eliminated. This series of improvement tests allowed the concrete to achieve a whiteness exceeding the international standard of 88%. It is the whitest fair-faced concrete building in China.

In recent years, China has witnessed rapid development under the strong leadership of the government. At the same time, a large number of commemorative, high-investment and

large-scale political landmarks have emerged. Most of them intervene into the city heroically, albeit somewhat regardless of urban context or of humanistic care, and many lack innovative architectural details. The cultural construction of the city urgently needs the support of private capital. China calls for buildings with innovative details, quality and integrity.

As the first art museum in Hunan Province to be fully operated by private capital, from design to end use, the museum has become an artistic benchmark, improving the standing of urban culture and changing the public's psychological recognition of architecture, reversing the strong political symbol of public buildings that is deeply rooted in the citizens' understanding. It has become the "city living room", the enlightenment base of art for children and the incubator cradle for local artists, art thus returns to life and returns to the public.

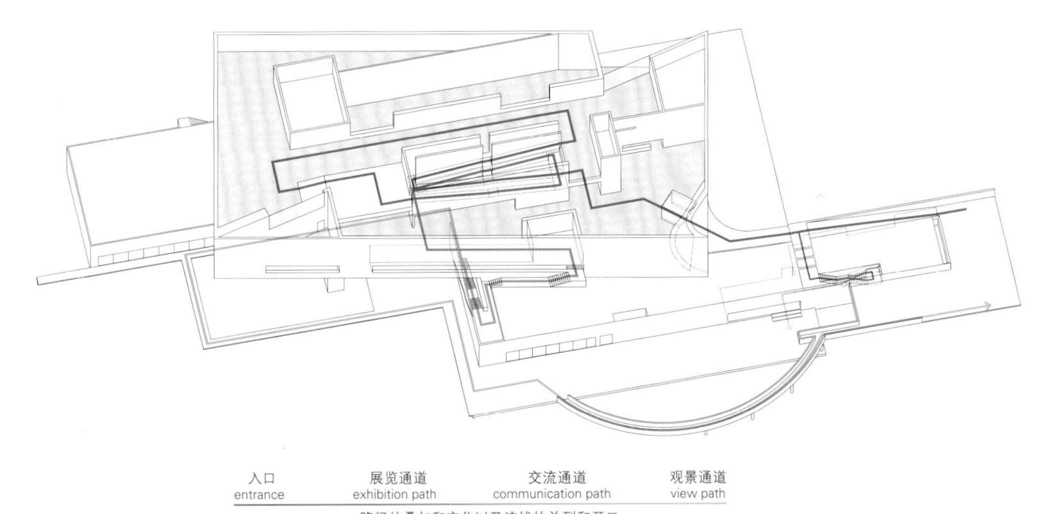

路径的叠加和变化以及流线的并列和开口
superposition and variation of the path and juxtaposition and opening of the streamline

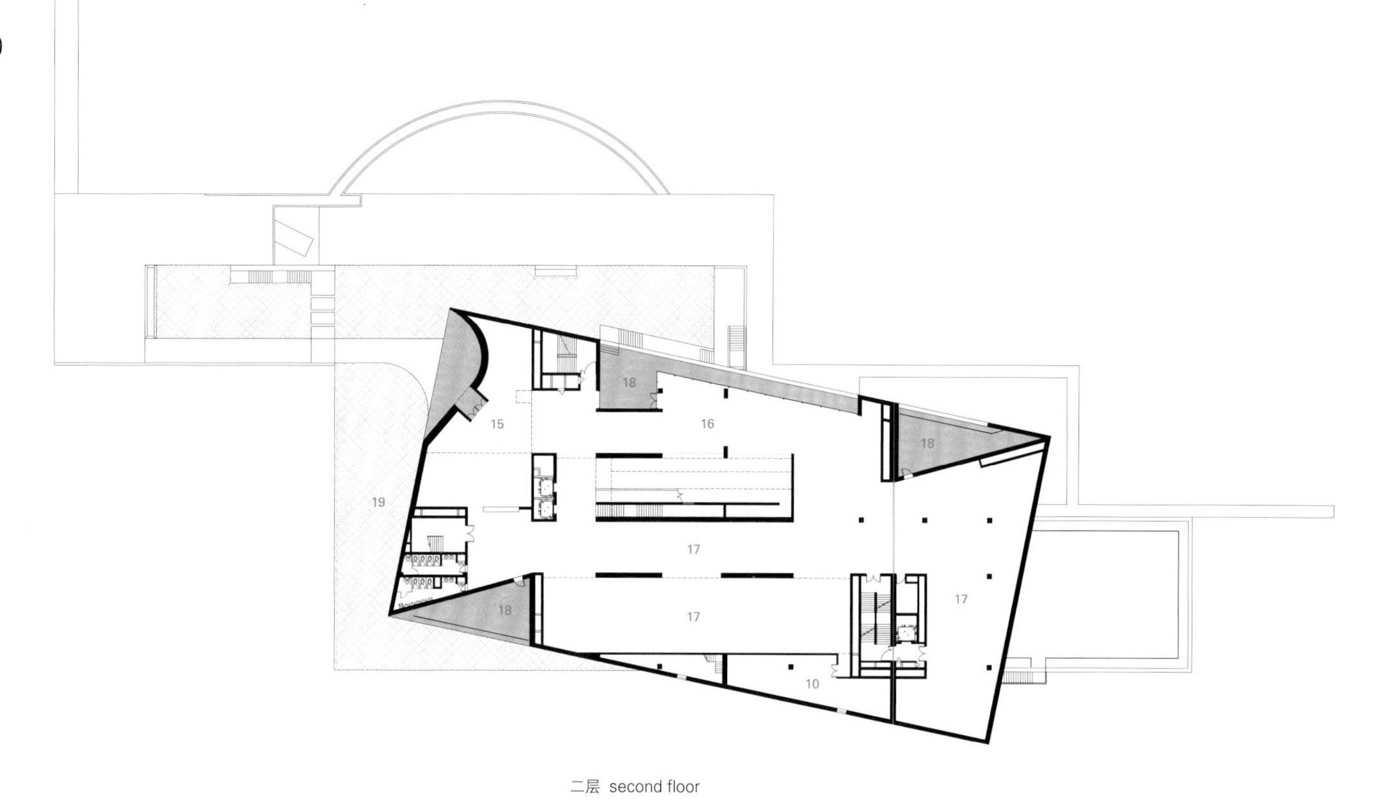

二层 second floor

1. 多功能间 2. 书吧、咖啡厅 3. 临时展厅 4. 日本料理 5. 停车场 6. 消防水池 7. 水泵房 8. 消防控制和智能机房 9. 配电室 10. 衣帽间 11. 空调机房 12. 户外广场、休息平台 13. 水面 14. 多功能服务大厅 15. 主入口大厅 16. 交易大厅 17. 专业展厅 18. 户外庭院 19. 无边界水面 20. 藏品库房 21. 摄影研究输出中心 22. 专业摄影工作室 23. 艺术工作室 24. 景观庭院 25. 入口大厅 26. 采光中庭 27. 屋檐下空间 28. 停车场

1. multifunction room 2. book bar, cafe 3. temporary exhibition hall 4. Japanese cuisine 5. parking garage 6. fire pool 7. water pump room 8. fire control room and intelligent machine room 9. electricity distribution room 10. locker room 11. air-conditioner engine room 12. outdoor square, rest platform 13. surface of the water 14. multi-functional service hall 15. main entrance hall 16. exchange hall 17. professional exhibition hall 18. outdoor courtyard 19. unbounded water surface 20. collection storeroom 21. photography research and output center 22. professional photography studio 23. artist studio 24. landscape courtyard 25. entrance lobby 26. light canyon 27. under the eaves space 28. parking garage

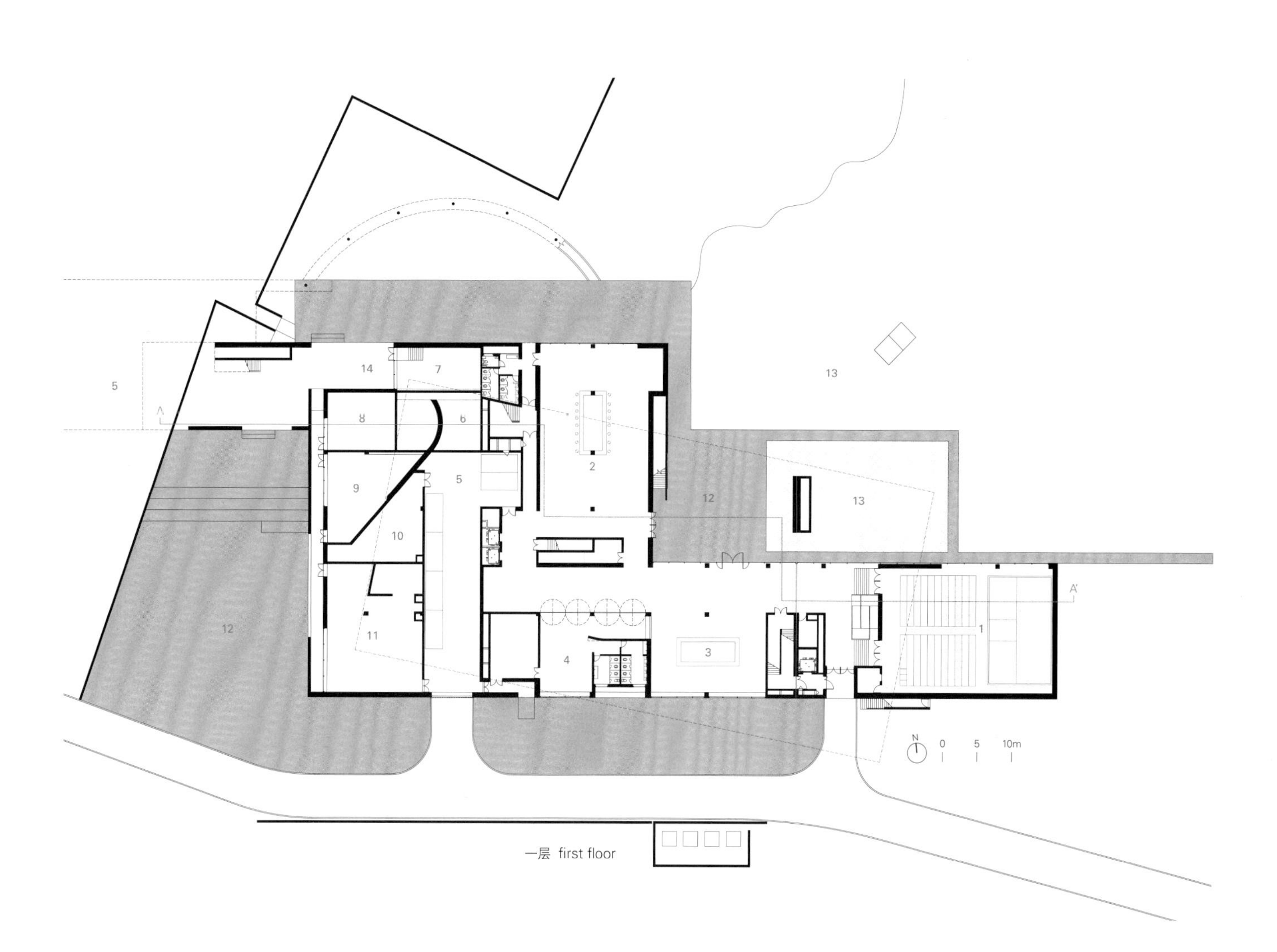

一层 first floor

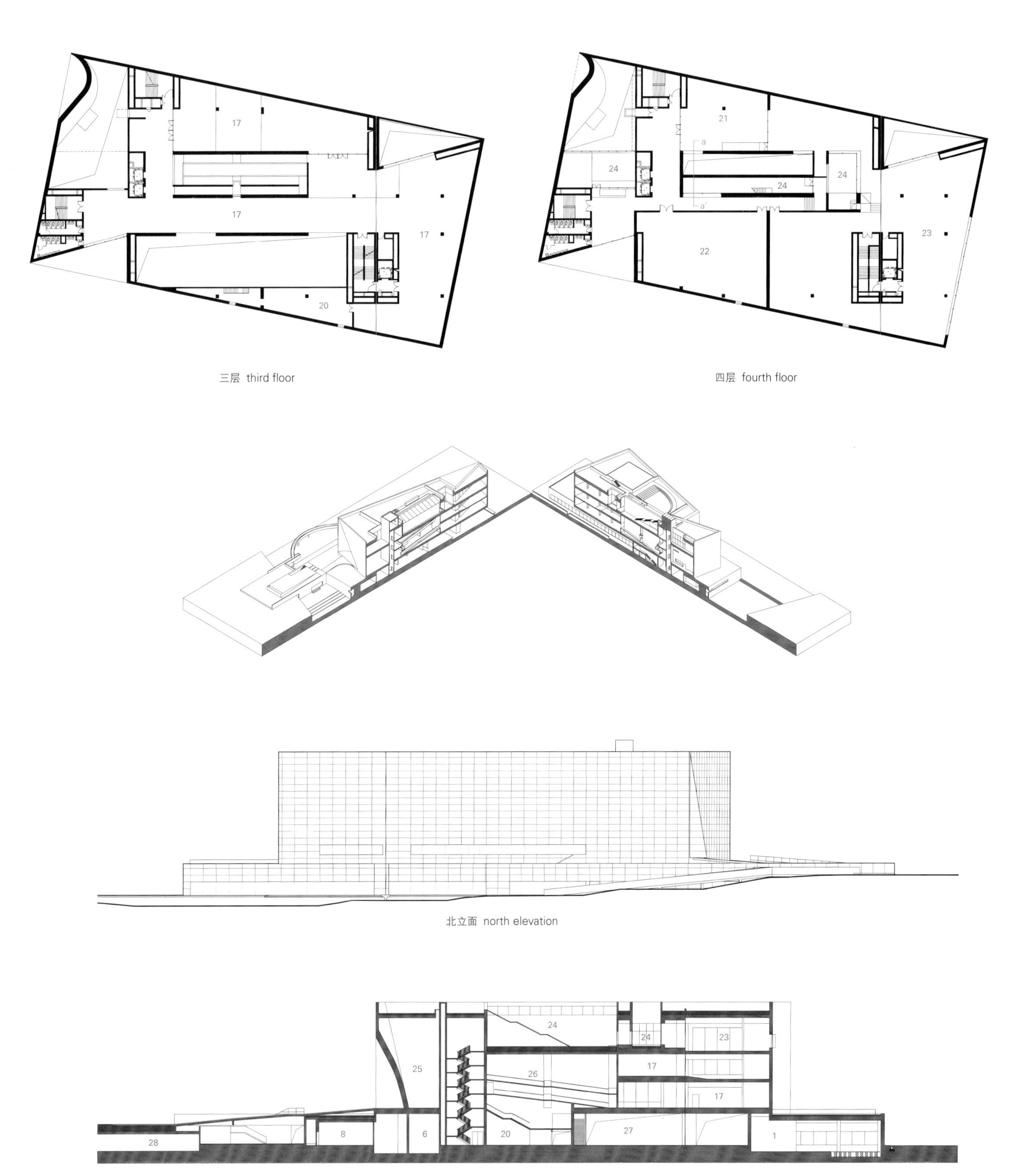

三层 third floor

四层 fourth floor

北立面 north elevation

A-A' 剖面图 section A-A'

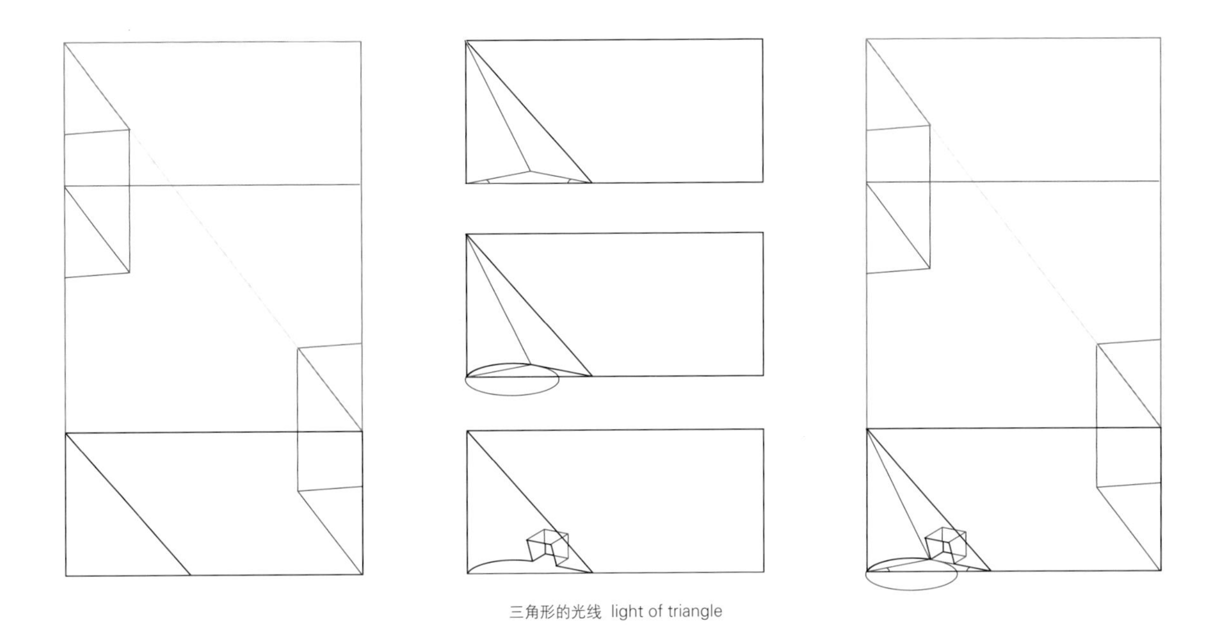

三角形的光线 light of triangle

项目名称：Xie Zilong Photography Museum / 地点：Changsha, China / 建筑师：WCY Regional Studio-Wei Chunyu, Zhang Guang / 设计团队：Shen Xin, Liu Haili, Chen Yun, Wen Yueming, Tong Chen, Chen Rongrong, Xiao Min (energy conservation), Zhu Jianhua (structural) / 室内设计：Li Xi, Li Jingbo / 机电设计：water supply and drainage-Zheng Shaoping, Liang Xiaoning; HVAC-Mao Yingjie, Zhang Ning; electric-Liu Jian / 景观设计：Hunan SLF Architecture and Landscape Design Co., Ltd.
照明设计：Wuhan Chenxin Technology Development Co., Ltd., Huizhou CDN Industrial Development Co., Ltd / 面积：10,621m²
设计时间：2015 / 竣工时间：2017 / 摄影师：©Yao Li (courtesy of the architect)

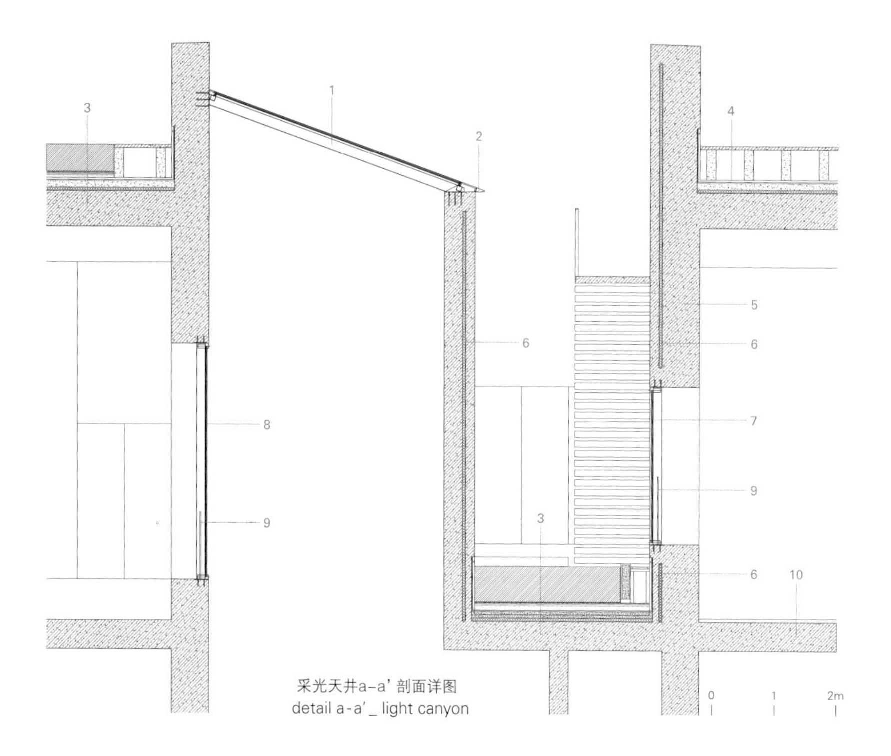

采光天井a–a' 剖面详图
detail a-a'_ light canyon

1. 160x88x6 I-beam
   8Low-E+12A+8+1.52PVB+8 toughened laminated insulating glass
2. 3 thick single layer gray fluorocarbon spraying aluminum plate
3. cast-in-place reinforced concrete floor slab
   20 thick 1:2.5 cement mortar screeding
   1.5 thick polyurethane waterproof coating vapor barrier,
   along the wall above the insulation
   surface 150 thick
   50 thick flame retardant extruded polystyrene board
   The thinnest 30 thick LC5.0 light aggregate concrete 1% looking for slope
   20 thick 1:3 cement mortar screeding
   2 thick synthetic polymer waterproof coating
   two layers 1.5 thick synthetic polymer waterproof rolling material
   4 thick SBS modified asphalt resistant to root puncture waterproof rolling materials
   20 thick 1:3 cement mortar protection layer
   high concave convex type (storage) water board
   geotextile filter layer
   150-550 thick modified soil or inorganic compound soil
4. 200x120x200 (h) C20 fine stone concrete block, 600-600, M5 cement mortar masonry
   The reinforced C25 water concrete precast plate 600x600x50
5. white fair-faced concrete wall
   transparent concrete protectant
6. thermal insulation board: 50 thick hard combustible extruded polyphenylene plate
7. 10 double silver low-e +12A+10 double ultra - white tempered insulating glass
8. 10 double silver low-e +12A+10 cesium potassium fire insulating glass
9. 12 thick super white toughened glass window panel
10. cast-in-place reinforced concrete floor slab
    grey fair-faced concrete ground

长江美术馆
Changjiang Art Museum

Vector Architects

美好生活综合服务商

一个以当代方式回应和纪念长江村生活的文化共享空间
A cultural and shared space establishes a contemporary response commemorating the life of the place

长江美术馆位于山西省太原市东北角的长江村。2016年，长江村像中国其他千万个村落一样，因城市建设的原因而被抹为平地，取而代之的是城市的高速发展。承载着人们日常生活的空间记忆，断裂成了碎片。长江美术馆作为一个未来服务于公众的文化共享空间，试图以当代的方式回应、纪念这片土地上人类建筑的痕迹和过往。

美术馆位于新建住宅社区南端边界，与城市紧邻。因此，如何使美术馆成为连接社区和城市的纽带，成为建筑师关注的主要问题之一。美术馆西南角的室外楼梯从城市街道开始向上爬升，穿过美术馆，通向二层的开放露台。这个露台成为一个被架起的广场，在庭院中央种了一棵树，这里也就成为公共活动的场所，再通过一条连桥延伸到街对面的北部社区。这条在外部交叉的交通流线完全独立于馆内的展览路线，广场和人行连桥共同服务于游客和当地居民。

在美术馆内部，一个底部直径5.7m、高16.4m的采光井是整体空间组织的"锚"。采光井既是起点也是终点。人们对这个空间的体验从底层开始，沿着自下而上的螺旋式楼梯，缓步直达画廊。在参观过程中，游客会不自觉地来到采光井旁，透过采光口回望各层，最终再沿着螺旋式楼梯自上而下，从位于采光井最高层的画廊移步至起点。在展览空间内，自然光经过天窗柔化过滤，再透过网格尺寸为1.9m的井字梁渗入展厅，形成一种匀质的、笼罩性的光线。

连接二层与四层的室外楼梯、四层突出的角窗、四层展厅西南角的竖向开窗，都为沉浸于展览当中的人们提供了回眸当今城市的瞬间。正如我们所看到的，长江美术馆就像一个砖砌的凝固体，承载着特定的时空坐标，默默而专注地见证了周围喧嚣的、瞬息万变的城市。

Changjiang Art Museum is located in Changjiang village at the northeast corner of Taiyuan, Shanxi Province. Similar to many other Chinese villages, this village was razed in 2016 to give way to full speed urban development. The spatial context that once inscribed the memories of people's everyday lives was disrupted and fractured. Changjiang Art Museum, as a cultural

and shared space that will serve the public in the future, attempts to establish a contemporary response commemorating the traces and atmosphere of the human construction that once existed on this piece of earth.

The museum is situated at the southern edge of a newly constructed residential community, adjacent to the urban gird. Therefore, how to make the museum function as a link between the community and the city became one of the architects' main concerns. At the lower level of the building, the designs carve out a space at the southwest corner for an outdoor staircase, beginning at the street level and leading up through the museum to an open terrace. The terrace at the second level becomes a raised plaza with a tree planted in the center of the courtyard, a space which allows for public activities and a further connection to the northern community across the street via a footbridge. This exterior crossing circula-

2003

2019

典雅的建筑 安放心灵的家

长江美术馆
CHANGJIANG ART MUSEUM

tion is public and independent from the route in the museum. Both the plaza and the footbridge accommodate uses for the museum's visitors and the local residents.

The galleries of the museum are arranged around a light well of 5.7m in diameter and 16.4m in height, which serves as the organizing "anchor" for all the spaces. The light well is both the starting point and the end point. People will start from passing through the bottom part of the light well and walk along the spiral staircase coiling up to the galleries while inadvertently looking back into the light well through the apertures at different levels, eventually concluding the journey by stepping down from the top gallery along the light well again back to the starting point. While inside the galleries, natural light is filtered and softened by the skylights, penetrating through a grid of 1.9 x 1.9m waffle beam and filling the interior with a homogeneous and immersive light quality.

Furthermore, the exterior staircase that connects the second through to the fourth floors, the protruding corner window on the fourth floor, and a southwest facing vertical window on the same floor, all provide museum visitors with glimpses of the contemporary cityscape of Taiyuan. As we see it, Changjiang Art Museum is like a solid block of brick – occupying a specific space-time coordinate, bearing witness silently and attentively to the clamorous and ever-changing city around it.

1. 门厅 2. 电梯厅 3. 物品寄放处 4. 卫生间 5. VIP室 6. 书店 7. 员工休息区 8. 储藏室 9. 礼品店 10. 庭院
11. 展厅 12. 露台 13. 采光中庭 14. 会议室 15. 开放办公室 16. 办公室 17. 接待处 18. 视频区

1. lobby 2. elevator hall 3. bag check 4. restroom 5. VIP room 6. book store 7. staff rest area 8. storage 9. goods lift hall
10. courtyard 11. exhibition hall 12. terrace 13. light atrium 14. meeting room 15. open office 16. office 17. reception 18. video area

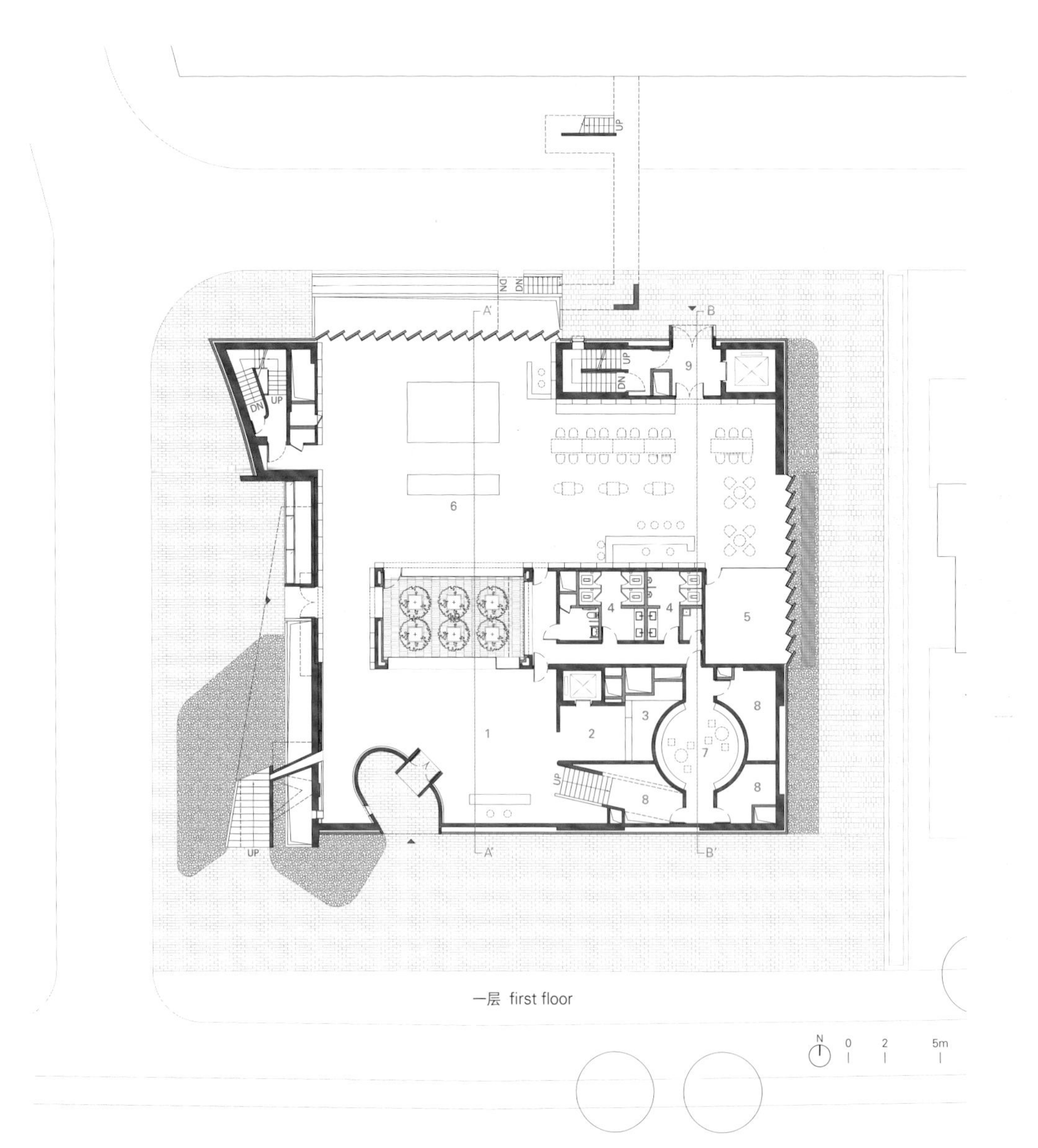

一层 first floor

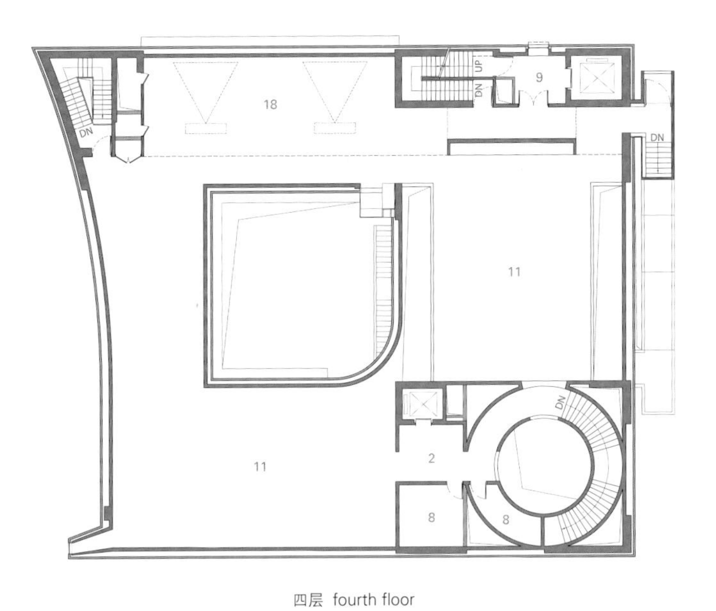

四层 fourth floor

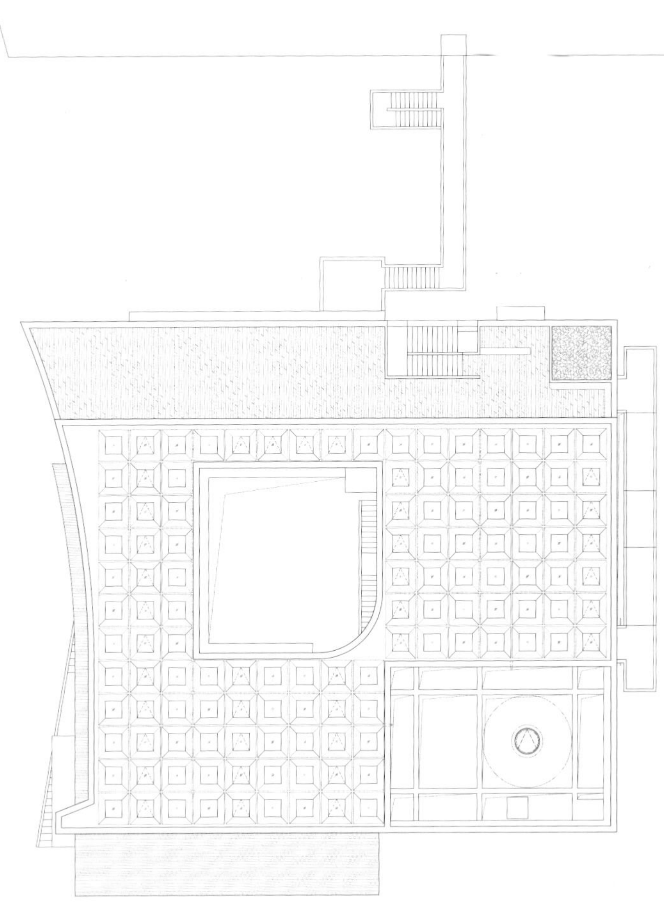

屋顶 roof

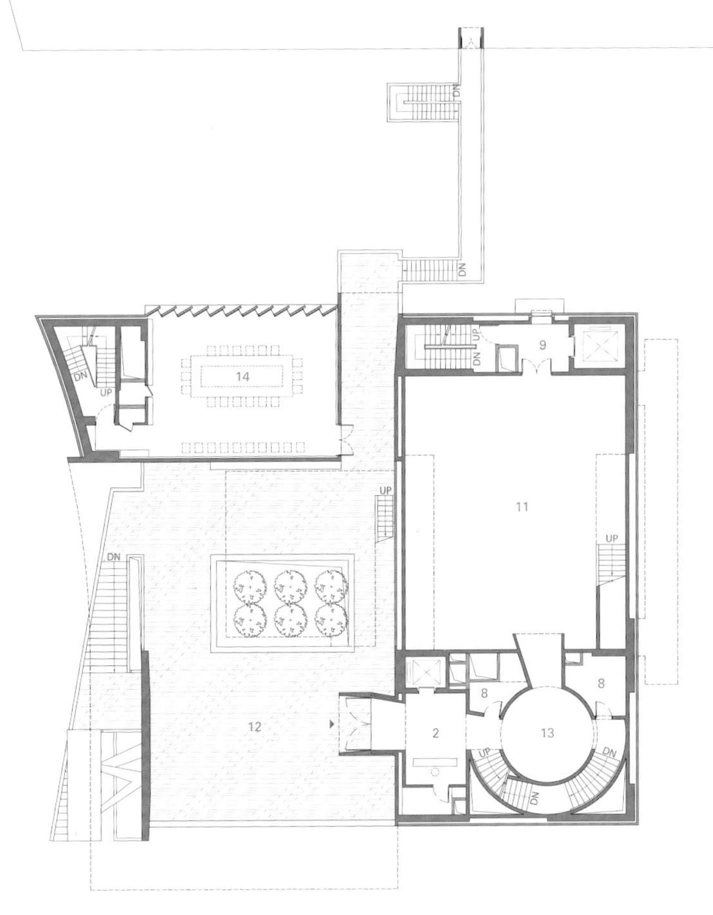

二层 second floor

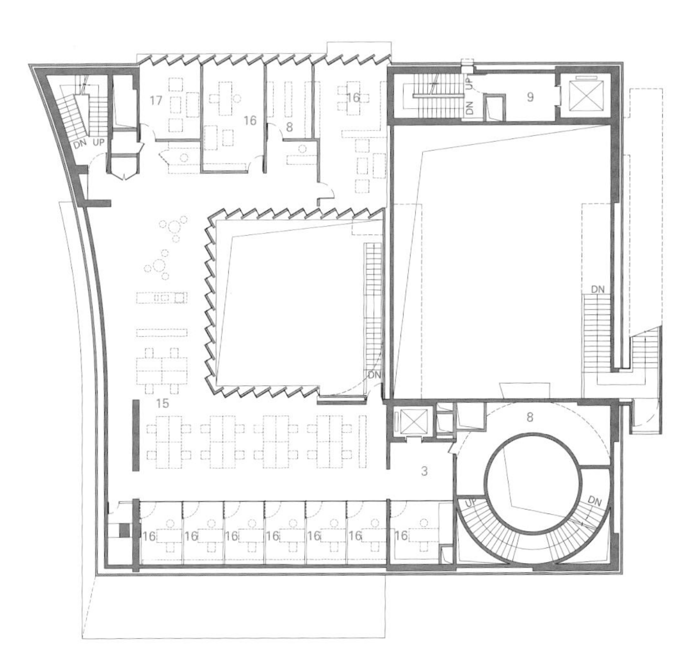

三层 third floor

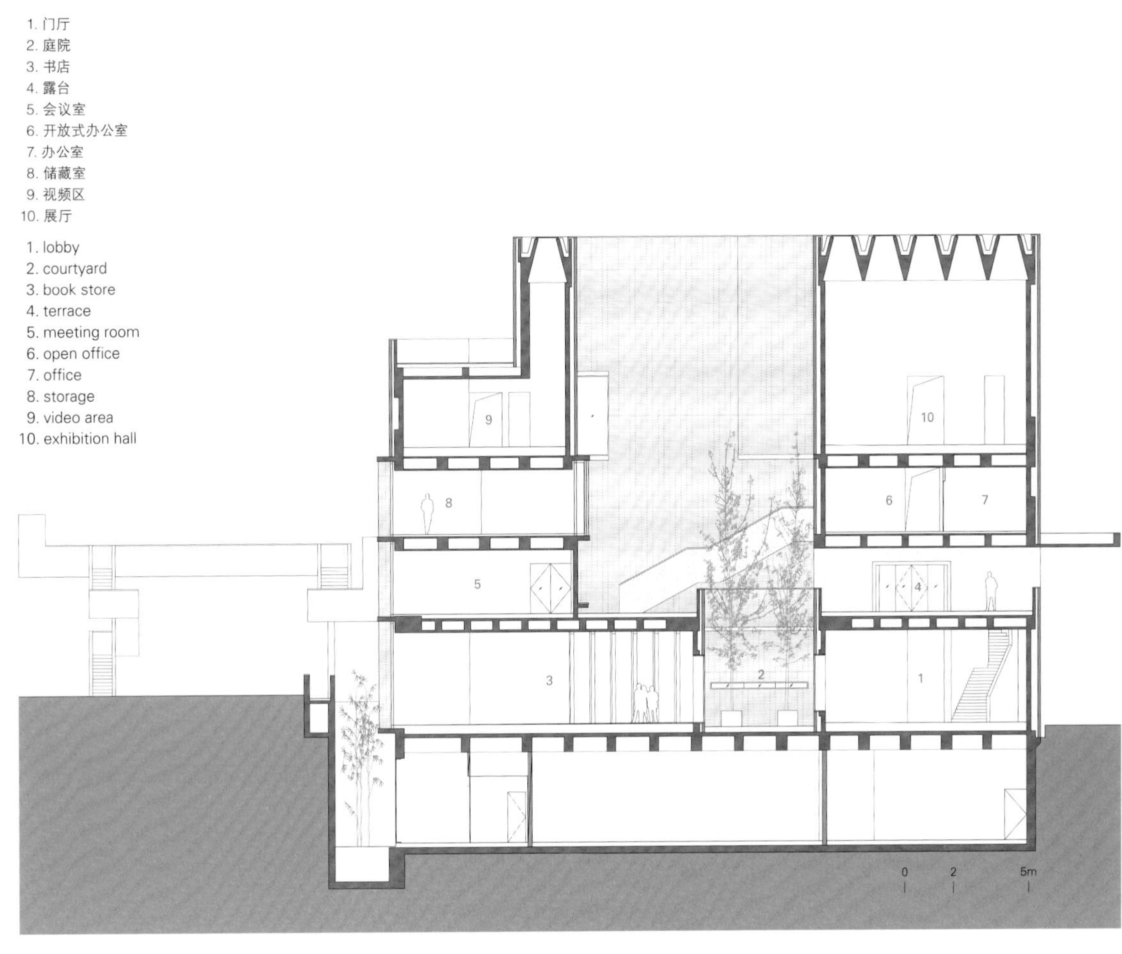

A–A' 剖面图 section A-A'

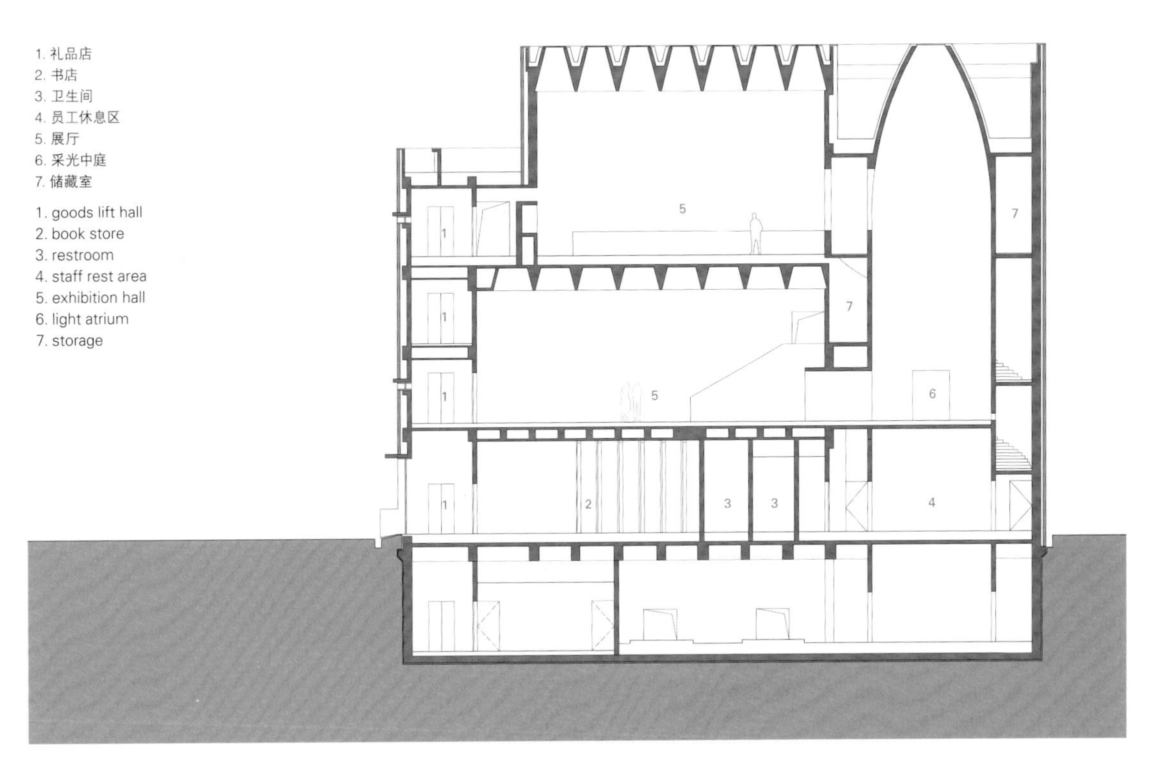

B–B' 剖面图 section B-B'

项目名称：Changjiang Art Museum / 地点：No.89, Kaixuan Road, Xinghualing District, Taiyuan, Shanxi, China / 建筑、室内、景观设计师：Vector Architects
设计主管：Dong Gong / 项目建筑师：Sun Dongping / 施工管理：Zhao Liangliang / 现场建筑师：Guo Tianshu, Chen Zhenqiang / 设计团队：Ma Xiaokai, Chen Zhenqiang, Zhang Kai, Jiang Yucheng, Teng Xiaotong, Zhao Dan / 结构&机电工程：Beijing Hongshi Design Co.,Ltd. / LDI项目建筑师：Zhang Cuizhen

LDI建筑师：Li Mo / 结构设计：Xue Wei, Zhong Zhihong, Tian Xi / 机电设计：Shen Juan, Hao Shufang, Zhang Jianxia, Shi Kefeng, Li Yingping / 照明顾问：X Studio, School of Architecture, Tsinghua University / 客户：Shanxi Qiandu Real Estate Development Co.,Ltd. / 建筑面积：3,932m² / 结构：reinforced concrete shear wall / 材料：wood-form concrete, brick, laminated bamboo / 设计时间：2016.5—2017.3 / 施工时间：2017.3—2019.9 / 摄影师：©Chen Hao (courtesy of the architect)

阿兰若艺术中心
Aranya Art Center

Neri & Hu Design and Research Office

沿着螺旋步道和圆形剧场形成的艺术空间为海滨度假社区打造了一个公共空间

The art space along spiral path and amphitheater provides a public space for the seaside resort community

Neri&Hu受富有进取意识的开发商阿兰若公司之邀，在其开发的海滨度假社区内设计一座艺术中心，Neri&Hu希望借此契机，尝试突破艺术中心与公共空间的固有设计概念。阿兰若社区非常强调生活方式的精神实质，即与环境的和谐互融。因而，尽管设计要求很简单，就是提供一个艺术中心，但设计方案是在内部庭院为居民创造一个公共空间，又可提供艺术品展示的空间。

当地海水受气候的影响，出现季节性的变化：夏季，海水蔚蓝而宁静；冬季则在表面结成了冰。艺术中心的概念设计便以季节性的海水变化为灵感，试图将大自然的奇妙囊括于建筑的内核。该方案最大限度地扩大了建筑占地面积，但在中心位置则挖凿出了一个纯粹的圆锥几何形状，在底部形成一座环形的阶梯式剧场。这种空间形式可以满足各种各样的使用需求，环形的广场在充满水时可以形成水景，排干净水之后又可作为表演和集会空间。上述展览空间得益于公共空间的整合，使得这个项目不仅仅是一个展示场所，同时也是一个社区分享的地点。

在厚重的建筑体量内有一系列互相联系的空间，游客可以在这里自由漫步，缓步向上，享受一段内外视野都很精彩的旅程。艺术空间的主要功能就是用来欣赏艺术的，阿兰若艺术中心也不例外。螺旋上升的步道将人们引导穿过所有的画廊空间，在展陈艺术品的过程中不断激发人们的观赏欲。倒锥形空间底部设置了一间咖啡馆、一间多功能画廊和一个圆形的室外剧场，从这里开始，环形连廊引领人们穿过五个独特的画廊空间，最终到达建筑顶部，在这里全方位地俯瞰四周的风景和建筑内部的活动。

建筑立面主要由不同纹理的混凝土砌块拼接而成，沉稳、厚重，有如一块坚韧的岩石，在不断变换的环境中屹立不动。光滑的表皮折射出万象的天空，而模块化的混凝土砖，则充分与日光互动，营造出另一种变幻的肌理。厚重的建筑立面使用青铜元素为点缀，捕捉了自然光，将行人的目光聚焦到每个画廊的入口。定制设计的照明与细节为朴实的色调增添了精致的细节。夜晚，室内的光线沿着倒锥形的空间散射出来，其光芒如同宝石，闪耀了整个滨海社区。

When the progressive developer Aranya asked Neri & Hu to design an art center inside their seaside resort community, Neri & Hu seized the opportunity to question the notions of space for art, versus communal space. Aranya, as a community, has a strong emphasis on the spiritual nature of their lifestyle ideology, a oneness with the environment. So, despite the straightforward brief to provide an art center, the design scheme is as much about creating a communal space for the residents, in the internal courtyard, as it is about the exhibition being displayed in the center.

Drawing inspiration from the seasonal ocean waters nearby – azure and calm in the summer, splintered ice throughout the winter – at its core, the building design attempts to encapsu-

late the natural wonder of water. The scheme maximizes its outer footprint but carves out a pure conical geometry at the center with a stepped amphitheater at the base. The central void space can be reconfigured and used in many ways, including a water feature when filled with water, but also a functional performance and gathering place when the water is drained. The exhibition galleries above benefit from the public space integration, making the project much more than just a place for display, but a place for sharing.

Within the thick mass of the building volume is a series of interlocking spaces that visitors can meander freely within, slowly ascending and enjoying a choreographed journey with directed views both inward and outward. Gallery spaces are about the enjoyment of art: this project is no different in that regard. A spiraling path leads the visitor through all the spaces, urging them onwards with the desire to see more. Starting at the bottom with the café, multi-purpose gallery, and an outdoor amphitheater, the path guides visitors through five distinct galleries, culminating at the rooftop where one gets a 360-degree view of the activities below.

Composed primarily of various textured concretes, with and without aggregate, the facade and materiality of the building are heavy in nature, like a solid rock sitting firmly in a shifting environment. Smooth surfaces reflect the changing skies, while the molded modular units pick up on the play of shadows throughout the day. Bronze elements act as accents on the heavy facade to catch light and draw attention to the entry of each gallery. Custom lighting and details add a touch of intricacy to the otherwise modest palette. In the evening, open modules allow light to shine through, the building is a jewel at the core of this seaside community.

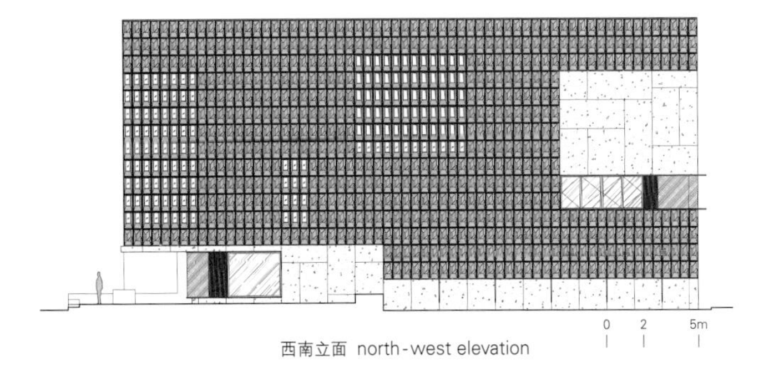

西南立面 north-west elevation

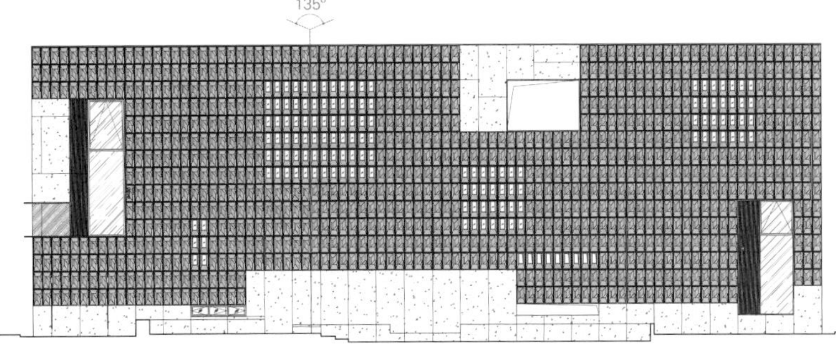

南立面 south elevation

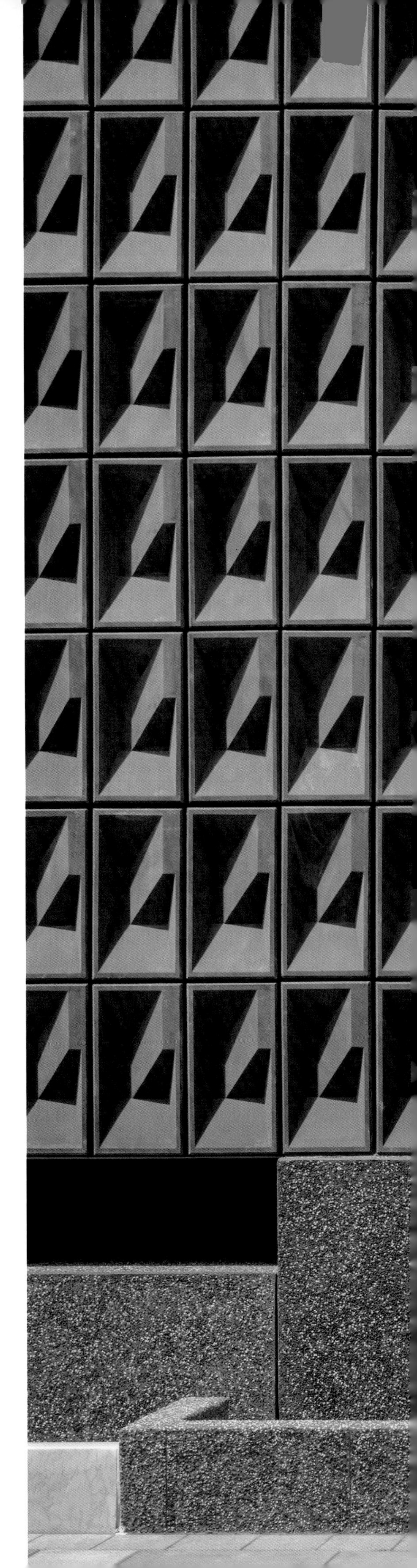

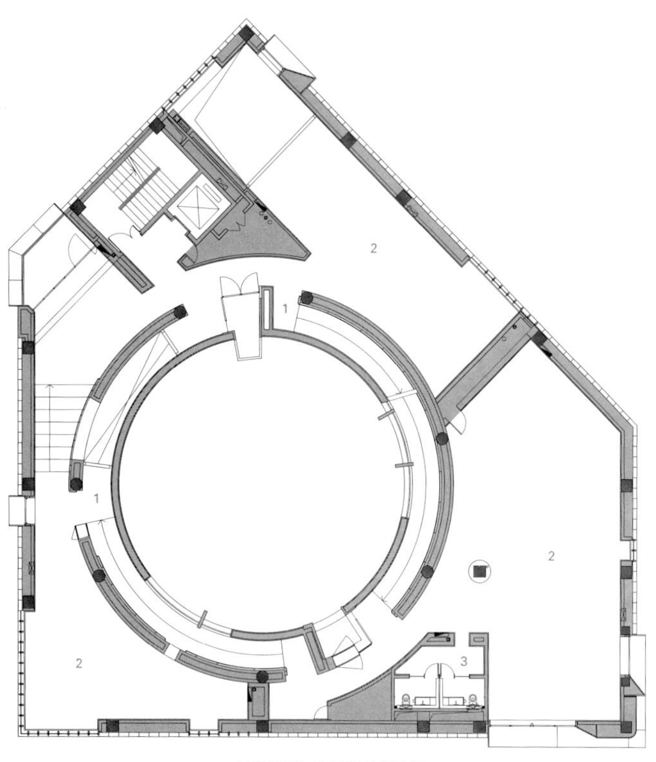

1. 环形坡道 2. 画廊 3. 卫生间
1. circular ramp 2. gallery 3. toilet
二层 second floor

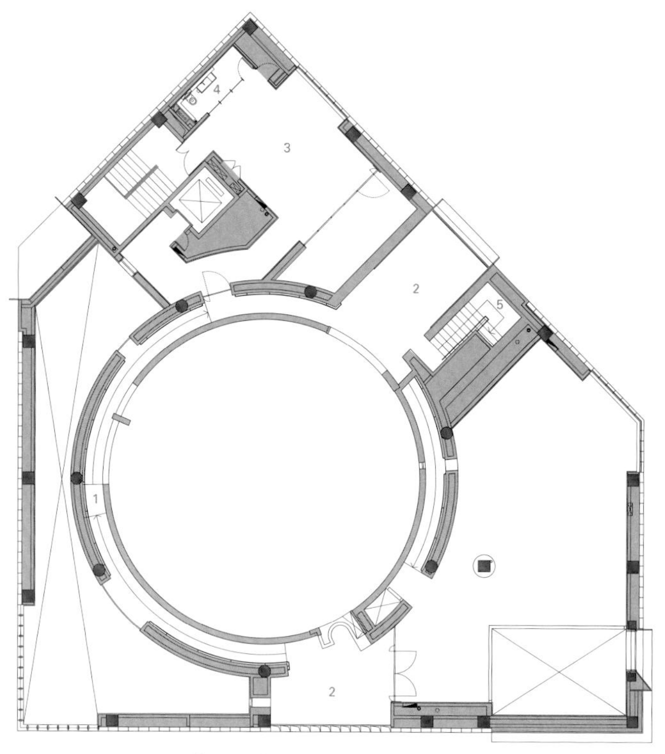

1. 环形坡道 2. 阳台 3. VIP室 4. 化妆室 5. 通往屋顶的楼梯
1. circular ramp 2. balcony 3. VIP room 4. powder room 5. stair to rooftop
三层 third floor

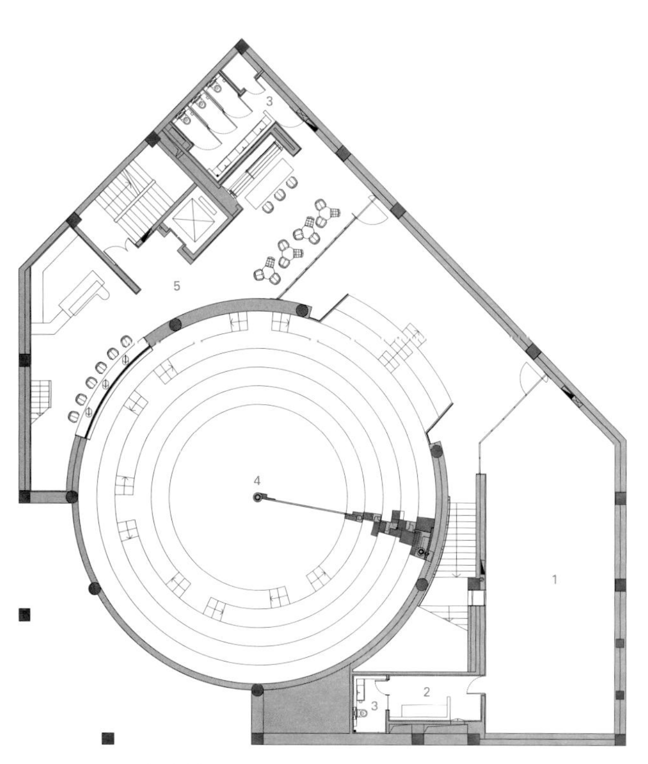

1. 多功能空间 2. 更衣室 3. 卫生间 4. 室外圆形剧场 5. 咖啡厅
1. multi-purpose space 2. changing room 3. toilet 4. outdoor amphitheater 5. café
地下一层 first floor below ground

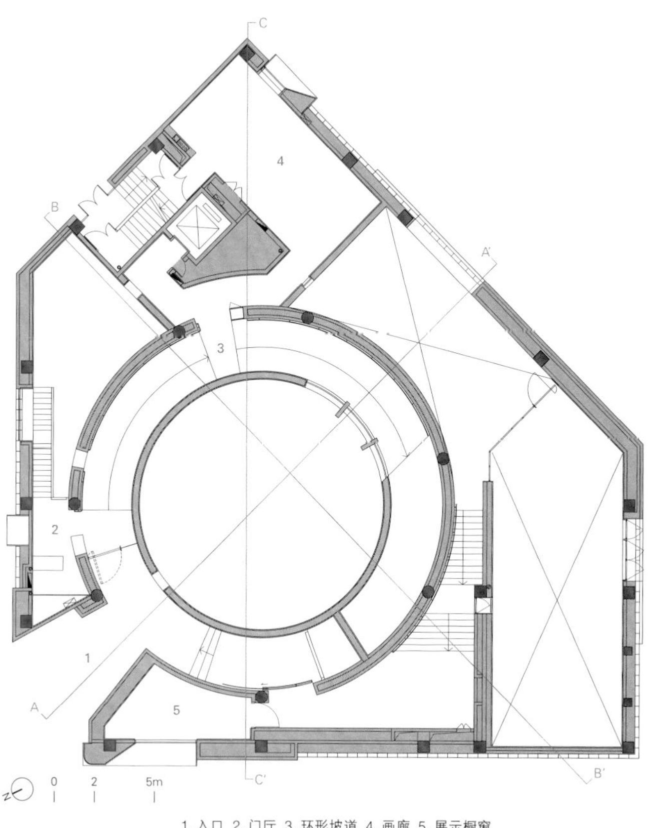

1. 入口 2. 门厅 3. 环形坡道 4. 画廊 5. 展示橱窗
1. entrance 2. foyer 3. circular ramp 4. gallery 5. display vitrine
一层 first floor

1. 室外圆形剧场
2. 入口
3. 环形坡道
4. 画廊
5. 阳台

1. outdoor amphitheater
2. entrance
3. circular ramp
4. gallery
5. balcony

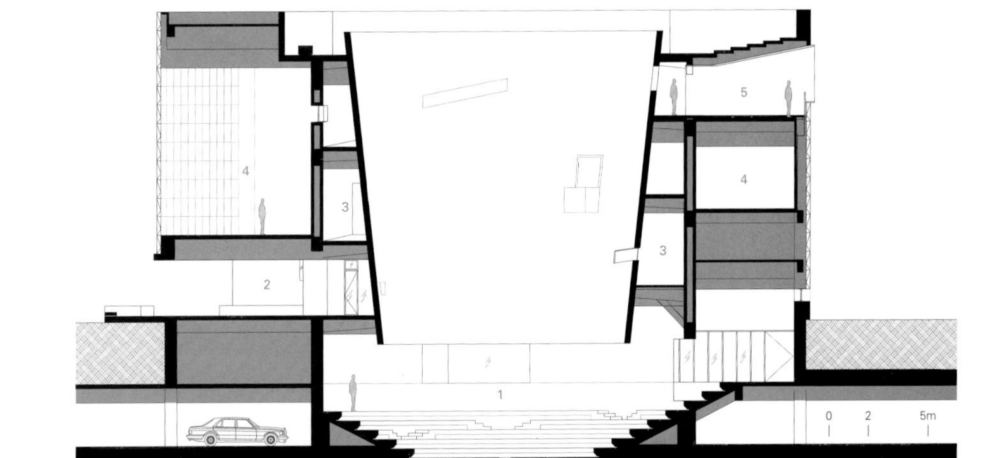

A-A' 剖面图 section A-A'

1. 室外圆形剧场
2. 咖啡厅
3. 环形坡道
4. 画廊

1. outdoor amphitheater
2. café
3. circular ramp
4. gallery

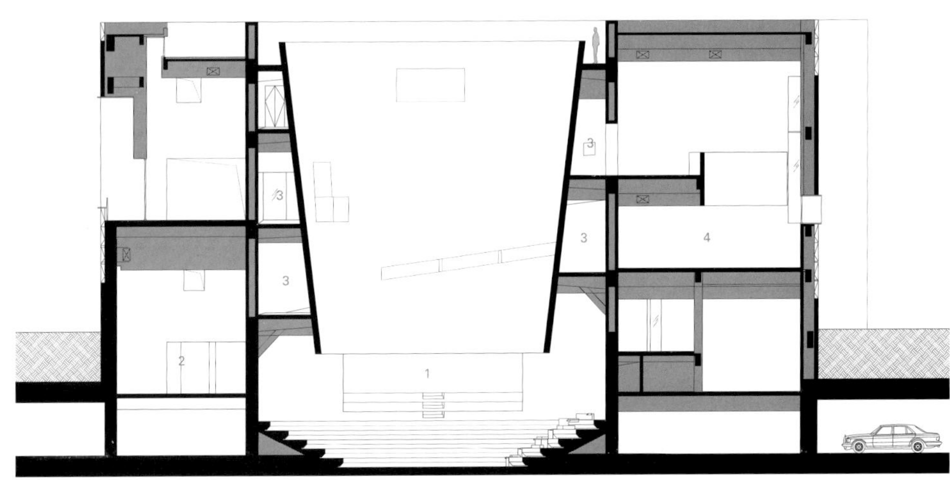
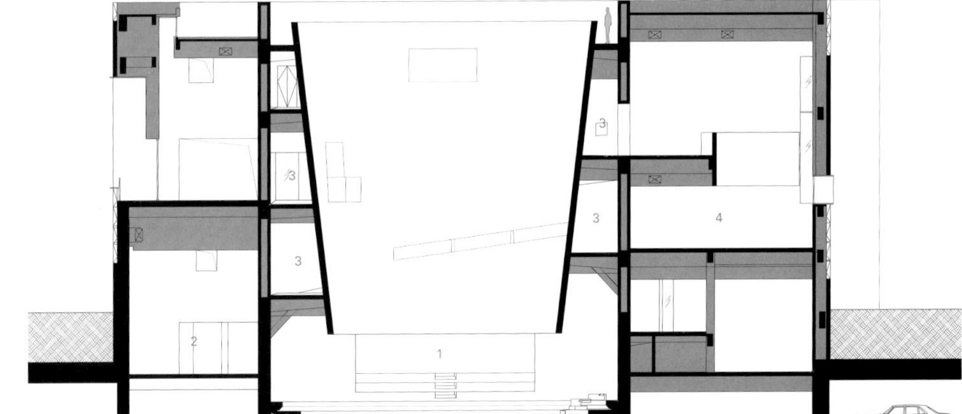

B-B' 剖面图 section B-B'

1. 室外圆形剧场
2. 咖啡厅
3. 环形坡道
4. 画廊
5. 展示橱窗
6. 阳台

1. outdoor amphitheater
2. café
3. circular ramp
4. gallery
5. display vitrine
6. balcony

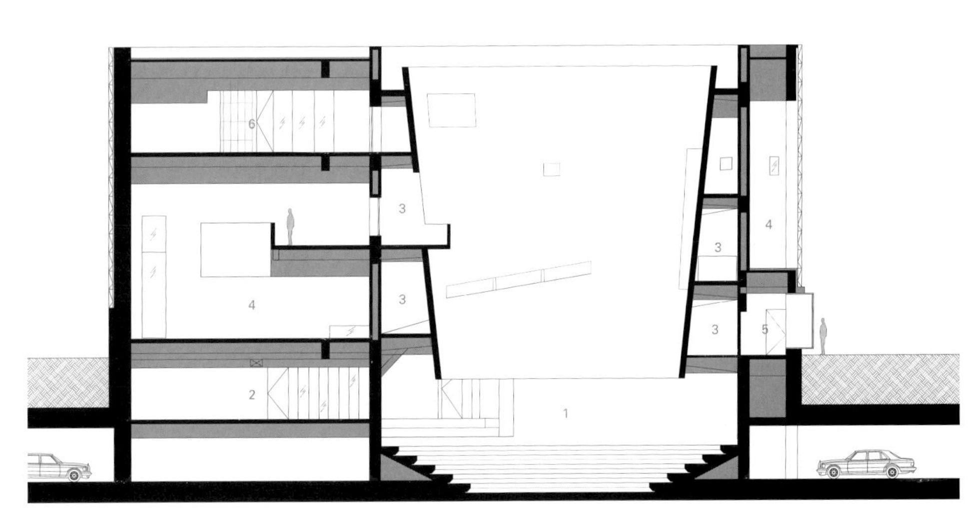

C-C' 剖面图 section C-C'

项目名称：Aranya Art Center / 地点：Block 4, South Zone, Aranya Golden Coast Community, Beidaihe New District, Qinhuangdao, Hebei, China / 建筑师&室内设计：Neri&Hu Design and Research Office / 创始合伙人、总负责人：Lyndon Neri, Rossana Hu / 设计团队：Nellie Yang (associate director, achitecture), Ellen Chen (associate & project manager) Jerry Guo (associate), Utsav Jain, Josh Murphy, Gianpaolo Taglietti, Zoe Gao, Susana Sanglas, Brian Lo (associate director, product design), Lili Cheng
总建筑面积：1,500m² / 功能：gallery, exhibition, multi-purpose space, café / 特别功能：outdoor amphitheater, modular concrete façade, circular courtyard with water feature, spiral ramp circulation / 材料：architectural-aggregate concrete panel, GFRC concrete module, sprayed concrete, bronze finish electroplated steel, bamboo wood doors; interiors-concrete, oak flooring, white paint, bronze, glass / 装修：Zuchetti, Duravit / 装饰照明：custom design / 家具：custom design, Stellar Works, Fritz Hansen
设计时间：2016.9—2019.5 / 摄影师：©Pedro Pegenaute (courtesy of the architect)

P200 Vector Architects

Gong Dong received B.Arch and M.Arch from Tsinghua University, followed by a diploma at the University of Illinois where he received the M.Arch. Also had an exchange experience at Technical University of Munich. He founded Vector Architects in 2008. Has won international awards such as 100+ Best Architecture Firms selected by *Domus* (2019); nominated for the Swiss Architectural Award (2018). Has been invited as a guest speaker by Universities and to a various major exhibitions, including the 2018 Venice Biennale. Has been teaching Design Studios at Tsinghua University since 2014.

P108 Vo Trong Nghia Architects

Vo Trong Nghia was born in Vietnam. He moved to Japan in 1996 as a government's scholarship student and started studying architecture. After graduation from Nagoya Institute of Technology in 2002, he joined the Department of Civil Engineering in the University of Tokyo and received Master's degree in 2004. In 2006, he started his firm in Ho Chi Minh City. Received international prizes and honors; ARCASIA Gold Medal and Building of the Year, FuturArc Green Leadership Award. Has continued to be involved in architecture by teaching at the Singapore University of Technology & Design in 2015. Is a visiting professor of Hiroshima Institute of Technology.

P160 Studio Zhu-Pei

Zhu Pei received his Master degree in Architecture both from Tsinghua University and UC Berkeley, he founded Studio Zhu-Pei in Beijing in 2005. Won The Architectural Review Award in 2017, Honor Award from the AIA New York Chapter in 2015. Has also been selected as an architecture jury member for Mies van der Rohe Award, Hong Kong Design Week, Korea International Competition in 2015, 2017 and IFLA Asia-Pac LA Awards in 2019. Has given many lectures at Harvard University, Columbia University, UC Berkeley, Tongji University, Tianjin University, etc.

©Ste Murray

P14 Grafton Architects

Yvonne Farrell[left] and Shelley McNamara[right] co-founded Grafton Architects in 1978 having graduated from University College Dublin in 1974. Teaching at the School of Architecture at University College Dublin from 1976 to 2002, they have been Visiting Professors at EPFL in 2010~2011. Currently, they are Professors at the Accademia di Architettura, Mendrisio, Switzerland. In 2018, they were the curators of the Venice Architecture Biennale. The practice has won numerous awards for their work. In 2016, they were honoured by being awarded the inaugural RIBA International Prize. Have also won the 2020 RIBA Royal Gold Medal and the 2020 Pritzker Prize. Directors Gerard Carty and Philippe O'Sullivan have been with the practice since 1992.

P152 Andrea Giannotti

Was born in Rome, graduated in Architecture at La Sapienza University in 2005. For multi-cultural exchange scenario he moved to Paris (2003-2005), again Rome (2005-2007), then Rotterdam (2007-2011), working for offices as EEA and OMA. Writes articles for *C3*, and for web magazines *Archdaily* and *The ArchHive*. Since 2011 is based in Beijing, working at Tsinghua University Architectural Design and Research Institute. In 2013 one of his projects has been awarded of the Human Habitat Prize; in 2016 one other of his projects has been awarded of the WA Experimental Design Award, and in 2017 exhibited in Florence and Beijing Design Weeks.

P102 Ana Souto

Graduated in Art History in Mardrid (2000), and complemented her studies with a PhD in Art History (Madrid, 2005) and in Hispanic and Latin American Studies (Nottingham, 2007). Is a Principal Lecturer of Architecture course at Nottingham Trent University, and supervises doctoral projects in the School of Architecture, Design and the Built Environment. Her research focuses on the analysis of the built environment as a repository of memory and identity. Has published articles in academic international journals.

P4 Aldo Vanini

Was born in Rome, used to practice in the fields of Architecture and Planning and now is retired. Was a member of regional and local government boards. He regularly publishes his works on international magazines and books. His principal fields of interest are now the relations between Architecture and Neurosciences, and the juridical aspects of Architecture, Landscape and Planning.

P4 Silvio Carta

Is an ARB RIBA architect and head of Design and Visual Arts, and chair of the Design Research Group at the University of Hertfordshire. His studies have focused on digital design, digital manufacturing, design informatics, data visualisation and computational optimisation of the design process. Silvio is an editor of A_MPS Architecture Media Politics and Society (UCL Press), and the curator of the international lecture series AUDITORIUM 2015-16: The Architecture of Information, Data, People and Public Space (Leuven, Belgium). Is head of the editorial board of *C3*, Korea.

P136 Barclay & Crousse Architecture

Was founded in 1994, by Sandra Barclay[left] and Jean Pierre Crousse[right] in Paris. They were born in Lima, Peru and trained as architects in Peru, Italy and France. Established their second office in Lima in 2006. Sandra is an Honorary Fellow of the American Institute of Architects. Jean Pierre is a director of the Master Program in Architecture at the Pontificia Universidad Católica del Perúand. Have earned the 2018 Mies Crown Hall Americas Prize, the 2016 Oscar Niemeyer Prize and a Special Mention at the 2016 Venice Biennale.

P58 Bez+Kock Architekten

Is a Stuttgart based office, founded in 2001 by Martin Bez[left] and Thorsten Kock[right]. Are a constantly growing team of about 40 architects. Martin Bez studied Architecture at the TU Karlsruhe and ETH Zurich (Swiss). Was deputy professor at the Technical University of Darmstadt in 2017~2018. Thorsten Kock studied Architecture at the University of Stuttgart and the Georgia Tech in Atlanta (USA). Is a member of the architectural advisory board for the City of Nurtingen. Their common goal is to realize unique, functional and economical buildings. One of our highest priorities is an integrated approach to the design process – from urban design to detail and from first stroke to structural realization.

P88 BIG

Was founded in 2005 by Bjarke Ingels, BIG is a Copenhagen, New York and London based group of architects, designers, urbanists, landscape professionals, interior and product designers, researchers, and inventors. Currently involves in a large number of projects throughout Europe, North America, Asia, and the Middle East. Believes that in order to deal with today's challenges, architecture can profitably move into a field that has been largely unexplored. A pragmatic utopian architecture that steers clear of the petrifying pragmatism of boring boxes and the naïve utopian ideas of digital formalism. Like a form of programmatic alchemist, it creates architecture by mixing conventional ingredients such as living, leisure, working, parking, and shopping. By hitting the fertile overlap between pragmatic and utopia, once again finds the freedom to change the surface of our planet, to better fit contemporary life forms.

P08 Douglas Murphy

Studied architecture at the Glasgow School of Art and the Royal College of Art, completing his studies in 2008. As a critic and historian, he is the author of *The Architecture of Failure* (Zero Books, 2009), on the legacy of 19th century iron and glass architecture, and *Last Futures* (Verso, 2015), on dreams of technology and nature in the 1960s and 70s. Is also an architecture correspondent for *Icon* magazine, and writes regularly for a wide range of publications on architecture and culture.

P52 Tom Van Malderen

His work ranges from buildings to objects, installations and exhibition design. His practice probes the intersections between art, design and architecture, and looks at the material gestures of everyday design and the construction of social space. After obtaining a Master in Architecture at Sint-Lucas Brussels, he worked with Atelier Lucien Kroll in Brussels and Architecture Project in London and Valletta. Since 2018, he works as an artist and exhibition designer based in Malta. He forms part of the contemporary art platform Fragmenta Malta and the Kinemastik international short film festival, and regularly contributes to international magazines. In Malta, his work is represented by the gallery Malta Contemporary Art.

Wei Chunyu

P182 WCY Regional Studio

Wei Chunyu is dean, professor, doctoral supervisor in School of Architecture, Hunan University; Founder of WCY Regional Studio; PhD in Architecture of Southeast University. Has studied traditional Chinese architecture and culture, exploring the boundaries in applying local materials and craftsmanship, focusing on the locality and difference of construction technique. Has been awarded the Chinese Architectural Education Award of the Chinese Architectural Society, the 2016 WA China Architectural Achievement Award, the 2019 ArchDaily Annual Chinese Architectural Award, and Gold Medal of the 2019 Architects Regional Council Asia Award.

P218 Neri & Hu Design and Research Office

Was founded in 2004 by Lyndon Neri[left] and Rossana Hu[right]. Is based in Shanghai, China with an additional office in London, UK. Neri received his B.Arch at the UC Berkeley and a M.Arch at Harvard. Was the Director for Projects in Asia and an Associate for Michael Graves & Associates in Princeton for over 10 years. Served as an active visiting critic at the Princeton, Syracuse, Harvard GSD, and UC Berkeley. Hu received her B.Arch from the UC Berkeley and a M.Arch at Princeton. Has worked for Michael Graves & Associates, Ralph Lerner Architect, SOM, and The Architects Collaborative (TAC). Has been a guest design critic at Princeton, UC Berkeley and Syracuse. Neri and Hu are the Overall Winner of The PLAN Award 2018. Also started teaching at Harvard Graduate School of Design the fall semester of 2019 as John C. Portman Design Critics in architecture.

©Andrew Rowat

Patkau Architects

P74 **Patkau Architects**

Was founded in 1978 by Patricia and John Patkau. John Patkau is licensed in a number of jurisdictions in both Canada and the United States. Holds a Master of Architecture degree from the University of Manitoba. Patricia Patkau holds a Master of Architecture degree from Yale University. She is a Professor Emerita in the School of Architecture at the University of British Columbia for 20 years. Received Prairie Wood Design Award of Excellence in Institutional Wood Design in 2020 and Canadian Wood Council Design Award.

P122 **Ignacio Urquiza**

Was born in Mexico City in 1983. Studied photography in Paris, France, graduated with an honorable mention in Architecture and Urbanism at Mexico City's Universidad Iberoamericana, and holds an MS in Advanced Architectural Design from Columbia University, New York. In 2008, Co-founded the Centro de Colaboración Arquitectónica and led the firm as Design Director until 2018. Establishes Ignacio Urquiza Arquitectos, an architecture studio based in Mexico City in 2019.

P122 **Bernardo Quinzaños**

Was born in Mexico City in 1984. Began his career in the visual arts as a painter and sculptor. Studied Architecture and Urbanism at the Universidad Iberoamericana in Mexico City and has participated in numerous architecture and art exhibitions both in Mexico and abroad. Is co-founder and Chief Executive Officer of CCA and Architecture Director of Grupo Invertierra.

P122 **Rodrigo Valenzuela Jerez**

Got a BA in Architecture and a MA in Visual Arts from the Universidad de Chile. Holds an MS in Advanced Architectural Design from Columbia University, New York. Is associate professor and coordinator of the design course at the Universidad de las Américas in Santiago, Chile. Developed EstudioRO–(E)Studio Futur@ and Rodrigo Valenzuela Jerez Arquitectos Asociados since 2014.

P122 **Camilo Moreno**

Graduated as an architect and has developed and coordinated architectural and urban projects in Mexico and Venezuela. Was joint director of Práctica de Arquitectura (PAR) from 2013 to 2015. Developed independent practice, Mexico City-based Lateral in 2016. In 2019 the practice name was changed to Colectivo Lateral de Arquitectura.

Urquiza

Bernardo Quinzaños

图书在版编目(CIP)数据

艺术空间的转变 / 丹麦BIG建筑事务所等编 ; 李璐译. — 大连 : 大连理工大学出版社, 2021.3
ISBN 978-7-5685-2940-2

Ⅰ. ①艺… Ⅱ. ①丹… ②李… Ⅲ. ①文化建筑－建筑设计 Ⅳ. ①TU242

中国版本图书馆CIP数据核字(2021)第018976号

出版发行：大连理工大学出版社
（地址：大连市软件园路80号　邮编：116023）
印　　刷：上海锦良印刷厂有限公司
幅面尺寸：225mm×300mm
印　　张：14.75
出版时间：2021年3月第1版
印刷时间：2021年3月第1次印刷
出 版 人：苏克治
统　　筹：房　磊
责任编辑：张昕焱
封面设计：王志峰
责任校对：杨　丹
书　　号：978-7-5685-2940-2
定　　价：298.00元

发　行：0411-84708842
传　真：0411-84701466
E-mail：12282980@qq.com
URL：http://dutp.dlut.edu.cn

本书如有印装质量问题，请与我社发行部联系更换。